Blockchain, Big Data and Machine Learning

Blockchain, Big Data and Machine Learning

Trends and Applications

Edited by

Neeraj Kumar

N. Gayathri

Md. Arafatur Rahman

B. Balamurugan

CRC Press
Taylor & Francis Group
Boca Raton London New York

CRC Press is an imprint of the
Taylor & Francis Group, an **informa** business

First edition published 2020
by CRC Press
6000 Broken Sound Parkway NW, Suite 300, Boca Raton, FL 33487-2742

and by CRC Press
2 Park Square, Milton Park, Abingdon, Oxon, OX14 4RN

© 2021 Taylor & Francis Group, LLC

First edition published by CRC Press 2020

CRC Press is an imprint of Taylor & Francis Group, LLC

ISBN: 9780367370688 (hbk)
ISBN: 978-0-367-37168-5 (pbk)
ISBN: 9780429352546 (ebk)

Typeset in Times
by Lumina Datamatics Limited

Contents

Preface

Blockchain is viewed as a revolutionary technology which addresses the modern techniques of decentralization, identity, data driven decisions, and data ownership. While blockchain applications and use cases are still in its nascent stages, the potential for digital transformation is immense. Blockchain influences Big Data and machine learning to find solution for storing and managing data in a distributed fashion on a P2P network. Blockchain technology can be a new part of the surrounding ecosystem of tools that Big Data uses. Actually, it can play a crucial role in security for user authentication, restricting access based on a user's need, recording data access histories, and proper use of encryption on data. Moreover, privacy and security are the two major concerns in Big Data analytics that should be significantly considered to safeguard the network. The main objective of this book is to explore threat intelligence, forensics and cyber security challenges in Big Data systems applying machine learning concepts. In particular, we are heading for providing solutions for the secured Big Data environment. This book is going to explicitly say all about the dead fusion in realizing the privacy and security of blockchain based data analytic environment.

Editors

Dr. Neeraj Kumar received his PhD in CSE from SMVD University, Katra (J & K), India, and was a postdoctoral research fellow in Coventry University, Coventry, UK. He is working as a Full professor in the department of computer science and engineering, Thapar Institute of Engineering & Technology, Patiala (Pb.), India since 2014. He has published more than 400 technical research papers, 10 edited/authored books, and 100 conferences publications with more than 10000 citations and h-index of 56. He has guided many research scholars leading to earn their PhD and ME/MTech degrees. His research is supported by funding from TCS, CSIT, UGC, and UGC in the area of smart grid, energy management, VANETs, and cloud computing. He is a member of the Cyber-Physical Systems and Security (CPSS) research group. He has research funding from DST, CSIR, UGC, and TCS of more than six crores from GOI. He has supervised 12 Ph.D students and guided 25 ME/MTech students with their thesis. He is a senior member of IEEE.

N. Gayathri received her ME degree in computer science and engineering. She is currently working as an assistant professor in School of Computing Science and Engineering at Galgotias University. She has published reputed 15 (SCI journals & Scopus indexed journals), and also presented papers in national/international conferences, published book chapters in CRC Press, IGI global, Springer, Elsevier and also have edited books in Wiley and CRC Press. Gayathri has six Indian patents. She is a guest editor for recent patents on computer science. Her research interests include Big Data analytics, IoT, and networks.

Md. Arafatur Rahman received his PhD degree in electronics and telecommunications engineering from the University of Naples Federico II, Naples, Italy in 2013. He has more than 10 years of research and teaching experience in the domain of computer and communications engineering. Currently, he is an associate professor with the Faculty of Computing, Universiti Malaysia Pahang, where he has conducted undergraduate and master's courses and supervised more than 21 BSc, 5 MSc, and 5 PhD students. He worked as a postdoctoral research fellow with University of Naples Federico II in 2014 and visiting researcher with the Sapienza University of Rome in 2016. His research interests include the Internet-of-Things (IoT), wireless communication networks, cognitive radio network, 5G, vehicular communication, cyber-physical systems, big data, cloud-fog-edge computing, machine learning, and security. He has developed an excellent track record of academic leadership as well as management and execution of international ICT projects that are supported by agencies in Italy, the EU, and Malaysia. Dr. Rahman has received number of prestigious international research awards, notably the Best Paper Award at ICNS 2015 (Italy), IC0902 Grant (France), Italian Government PhD Research Scholarship and IIUM Best Masters Student Award, Best Supervisor Award at UMP, Awards in International Exhibitions including Diamond and Gold in BiS 2017 UK, Best of the Best Innovation Award and Most Commercial IT Innovation Award, Malaysia, and Gold and Silver medals in iENA 2017 Germany. Dr. Rahman has co-authored articles in around 100 prestigious IEEE and Elsevier journals (e.g., *IEEE TII, IEEE TSC, IEEE COMMAG, Elsevier JNCA*, and *Elsevier FGCS*) and conference publications (e.g., *IEEE Globecom, IEEE DASC*) and has served as an advisory board member, editor (Computers, MDPI), lead guest editor (IEEE ACCESS, Computers), associate editor (IEEE ACCESS), chair, publicity chair, session chair, and member of Technical Programme Committee (TPC) in numerous leading conferences worldwide (e.g., IEEE Globecom, IEEE DASC, IEEE iSCI) and journal editorial boards. He is a fellow of IBM Center of Excellence and Earth Resources & Sustainability Center, Malaysia and a senior member of IEEE.

Dr. B. Balamurugan completed PhD at VIT University, Vellore and is currently working as a professor in Galgotias University, Greater Noida, Uttar Pradesh. He has 15 years of teaching experience in the field of computer science. His area of interest lies in the field of IoT, Big Data, and networking. He has published more than 100 international journal papers and contributed book chapters.

Contributors

Khushbu Agrawal
School of Computing Science and
    Engineering
Department of Computer Science
Galgotias University
Greater Noida, India

Rashid Ali
Department of Computer Engineering
Aligarh Muslim University
Aligarh, India

P. Balakrishnan
SCOPE, VIT University
Vellore, India

S. Balakrishnan
Department of Computer Science and
    Business Systems
Sri Krishna College of Engineering and
    Technology
Coimbatore, India

B. Balamurugan
School of Computing Science and
    Engineering
Galgotias University
Greater Noida, India

Syed Muzamil Basha
Department of Information Technology
Sri Krishna College of Engineering and
    Technology
Coimbatore, India

S. Dhivya
Department of Computer Science and
    Engineering
Sri Shakthi Institute of Engineering and
    Technology
Coimbatore, India

S. V. Evangelin Sonia
Department of Computer Science and
    Engineering
Sri Shakthi Institute of Engineering and
    Technology
Coimbatore, India

N. Gayathri
School of Computing Science and
    Engineering
Galgotias University
Greater Noida, India

L. Godlin Atlas
Department of Computer Science and
    Engineering
Sri Shakthi Institute of Engineering and
    Technology
Coimbatore, India

P. Hamsagayathri
Department of Electronics and
    Communication Engineering
Bannari Amman Institute of
    Technology
Erode, India

R. Indrakumari
School of Computing Science and
    Engineering
Galgotias University
Greater Noida, India

J. Janet
Department of Computer Science and
    Engineering
Sri Krishna College of Engineering and
    Technology
Coimbatore, India

Suresh Kallam
Department of Computer Science and
    Engineering
Sree Vidyanikethan Engineering
    College
Andhra Pradesh, India

Firoz Khan
Higher Colleges of Technology
Abu Dhabi, United Arab Emirates

Abhishek Kumar
Department of Computer Science
Banaras Hindu University
Varanasi, India

Harsh Kumar
School of Computing Science and
    Engineering
Department of Computer Science
Galgotias University
Greater Noida, India

Neeraj Kumar
Department of Computer Science and
    Engineering
Thapar Institute of Engineering and
    Technology
Patiala, India

C. Mageshkumar
Department of Computer Science and
    Engineering
Sri Shakthi Institute of Engineering and
    Technology
Coimbatore, India

Basetty Mallikarjuna
School of Computing Science and
    Engineering
Galgotias University
Greater Noida, India

R. Manikandan
School of Computing
SASTRA Deemed University
Thanjavur, India

M. R. Manu
School of Computing Science and
    Engineering
Department of Computer Science
Galgotias University
Greater Noida, India

Anurag Pandey
Department of Computer Science
Banaras Hindu University
Varanasi, India

Rizwan Patan
Department of Computer Science and
    Engineering
Velagapudi Ramakrishna Siddhartha
    Engineering College
Vijayawada, India
and
Sree Vidyanikethan Engineering
    College
Tirupati, India

Robbi Rahim
Department of Management
Sekolah Tinggi Ilmu Manajemen
    Sukma
Medan, Indonesia

Md. Arafatur Rahman
Faculty of Computer System and
    Software Engineering
University Malaysia Pahang
Pahang, Malaysia

K. Rajakumari
Department of Computer Science and
    Engineering
Sri Shakthi Institute of Engineering
Coimbatore, India

S. Rakesh Kumar
School of Computing Science and
    Engineering
Galgotias University
Greater Noida, India

T. V. Ramana
Federal Technical and Vocational
Education and Training Institute
   (FTVET)
Addis Ababa, Ethiopia

L. Ramanathan
SCOPE, VIT University
Vellore, India

P. Ramya
Bannari Amman Institute of
   Technology
Department of Electronics and
   Communication Engineering
Erode, India

Tawseef Ayoub Shaikh
Department of Computer Engineering
Aligarh Muslim University
Aligarh, India

Yogesh Sharma
Maharaja Agrasen Institute of
   Technology
Rohini, Delhi

K. Shoukath Ali
Bannari Amman Institute of
   Technology
Erode, India

Arnab Kumar Show
Department of Computer Science
Banaras Hindu University
Varanasi, India

Achintya Singhal
Department of Computer Science
Banaras Hindu University
Varanasi, India

P. Suvithavani
Department of Computer Science and
   Engineering
Sri Shakthi Institute of Engineering and
   Technology
Coimbatore, India

R. Venkatesh
Department of Computer Science
Sri Shakthi Institute of Engineering and
   Technology
Coimbatore, India

1 Introduction to Blockchain and Big Data

Robbi Rahim, Rizwan Patan,
R. Manikandan, and S. Rakesh Kumar

CONTENTS

1.1 BLOCKCHAIN BASIC TECHNOLOGIES

The first journey of blockchain technology was Bitcoin, It was a form of cryptocurrency designed by Nakamoto to design a swift, an inexpensive and translucent peer-to-peer money transaction. With the acceleration and speedy movement of internet era, the future industrial revolution also demanded for the requirement of to improve the privacy of data-driven enterprise architecture, including but not limited to decentralization, persistency, anonymity, and auditability.

A blockchain [1] consisting of two words block and chain refers to a continuous growing list digital record in the form of packets also called as blocks that are linked and secured with the aid of cryptographic mechanism. The blockchain, also referred to as the digitally recorded blocks of data are secured and stored in the form of a linear chain. Each block in the linear chain comprises of several data, i.e., Bitcoin transaction.

This Bitcoin transaction on the other hand is secured via cryptographically hashed following time stamped technology. When a new block is formed, it will contain a hash of the previous block. These blocks are chronologically ordered initiating from the first block since the inception in the entire blockchain to the newly formed block. This process is repeated until it grows and maintains the network.

1.1.1 BASICS OF BLOCKCHAIN AND ITS ARCHITECTURE

A blockchain simply is referred to as the chain of blocks in a digital format. It is also referred to as the decentralized ledger that records all transactions. The blockchain has already been utilized for management of individual identity by several researchers in the field of research community. However, a new set of regulations has been brought into by several researchers while dealing with the personal information of users concerning blockchain [2].

Each and every time whenever a user or customer make a purchase of digital coins via decentralized exchange, sells or transfers coins, a digital ledger records that specific transaction in an encrypted format, not understandable to anyone. In this way, without the need of the third party, the transaction recorded in digital format is said to be safeguarded from cybercriminals. The figurative representation of blockchain is shown in Figure 1.1.

As shown in Figure 1.1, four blocks are included with each block connected in chain with the other blocks. Here, block 1 is linked to block 2, block 2 is lined to block 3, block 3 is linked to block 4 and block 4 is linked to block 1, forming a chain. As illustrated in this figure, the block in other words is considered as the container for data, where the data is said to be stored in that specific container. The structure of a block is given below.

As far as Bitcoin blockchain, each block is composed of data referring to the Bitcoin transactions, Block Header, Block Identifies, and Merkle Trees. This section provides a detail description of the block structure (Table 1.1) [3]. Some of the normally used idioms of blockchain technologies involving the design of blockchain architecture [4] are shown below in Figure 1.2 along with the description.

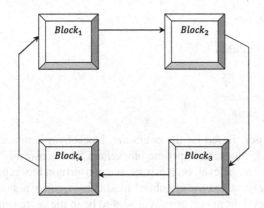

FIGURE 1.1 A sample blockchain.

TABLE 1.1
Block Structure

Field	Description	Size
Block size	Size of block in bytes	4 bytes
Block header	Several fields comprises of block header	80 byes
Transaction counters	Frequency of transactions	1–9 bytes
Transaction	Actual transactions recorded	Variable

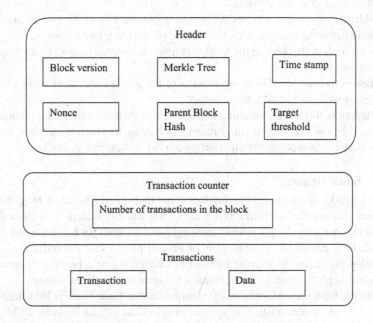

FIGURE 1.2 Blockchain architecture.

- Blockchain types
- Node
- Consensus algorithm
- Block
- Header
- Transaction counter
- Transaction data

Blockchain types: Based on the operation, blockchain is divided into three types—public blockchain, private blockchain, and consortium blockchain. In the organization level, both private and consortium blockchain are used. On the other hand, in case of public blockchain, security is said to be in the increasing level; however, privacy is said to be in the decreasing level.

Node: Node is alternatively represented as a computer. It is said to be possessed by an organization participating in the blockchain network. It is also referred to as a user. Node remains the central owner of any blockchain. Its task remains in verifying the transactions with other nodes or computer. The node in the blockchain framework forms as the association point between blockchain technology and user.

Consensus algorithm: The consensus algorithm is also referred to as the agreement between the blockchain and the user. It is used in the framework with the purpose of approving the decisions for nodes or machines. Some of the familiar consensus algorithm is proof of work (PoW), proof of stake (PoS), proof of burn (PoB), and so on.

Block: A block is referred to as the transaction decision included in the current chain after effective consent.

Header: Block version denotes the presently obtained block version in blockchain network. In the header, besides the predecessor block hash, successor block hash is also kept in the header. A nonce is utilized to modify the block hash output.

Transaction counter: The serial number of the present and previous block is represented via a transaction counter.

Transaction data: The meaning of this field, i.e., transaction data changes based on the usability. It either refers to a Bitcoin transaction, records, user personal information, information pertaining to healthcare, and so on.

1.1.1.1 Block Header

The block header is composed of metadata about that specific block. A block header is utilized to recognize a specific block on a complete blockchain. It is hashed in a repeated manner with the purpose of creating proof of work for mining rewards. As the blockchain comprises of sequences of several blocks that are utilized to store information pertaining to specific transactions occurring on a blockchain network, with the aid of block header, differentiation between the blocks are made.

The blockchain network consists of a unique header. Each block is identified with the aid of block header hash. The specifications of the header include, an 80-byte long string supporting a 4-byte long Bitcoin version number with a 32-byte previous block hash. It also includes a 32-byte long Merkle root supporting a 4-byte long time-stamp with a 4-byte long nonce utilized by the cryptography miners. Along with the above description, the block header includes the following:

- Cryptographic hash
- Mining competition
- Data structure to summarize the transactions in the block

1.1.1.2 Block Identifiers

The block identifiers are specifically the cryptographic hash. With these block identifiers, the specific block is said to be identified in a unique manner. Usually, two block identifiers are said to exist—block header hash and block height. As far as block header hash are concerned, the block's hash is evaluated by each node as soon as the block is received from the network. The second method to obtain a block is via its position in the blockchain, which is simply called as the block height.

1.1.1.3 Merkle Trees

The Merkle tree refers to the framework of transactions in the specific block for the corresponding blockchain. The Merkle tree is utilized to store the summary of all transactions in the block and refers to a data structure utilized for summarizing and verifying the dataset integrity containing cryptographic hashes. It is also a method to summarize the entire transactions pertaining to a block, ensuring an efficient process to verify whether a transaction is included in a block or not. Figure 1.3 given below illustrates a sample blockchain Merkle tree.

1.1.1.4 Features of Blockchain

Comprehending how a blockchain functions [5] from technical aspect, the characteristics of blockchain should be understood. There are several things that the blockchain is ready for change. It is reasonable that the Blockchain can really damage multiple institutions and organizations and make the operations more representative, secure, unambiguous, and effective. Blockchain steers an indispensable and disorderly tendency in several fields and aspects. With the growth in the evolution, several intermediaries reduces in number, hence the overall system forms a decentralization form, ensuring security and transparency.

Some of the key characteristics of blockchain is listed below is an adaptable one, in that sense it is said to be altered to fit specific purposes. Figure 1.4 given below shows the blockchain features, followed by which the description about the features are provided.

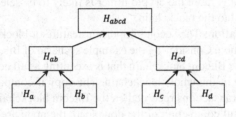

FIGURE 1.3 Calculating the blocks using Merkle tree.

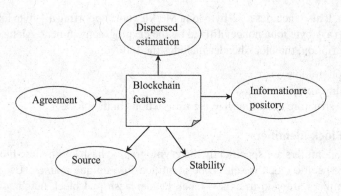

FIGURE 1.4   Blockchain features.

As shown in Figure 1.4, the blockchain features include:

- Agreement
- Dispersed estimation
- Information repository
- Source
- Stability

The descriptive nature of blockchian features are elaborated as follows.

Agreement: The first and foremost feature of blockchain is the agreement. Agreement, as the terms itself defines the potentiality of all the anonymous network users, agreeing the network's rules and regulations. Besides, whether the rules and regulations are being following by the network users or not are also monitored in a regular basis in the blockchain framework. The agreement in blockchain is said to be arrived at in several means like PoW algorithm or PoS algorithm and so on.

As far as the PoW algorithm is concerned, complex mathematical equations are solved by crypto currency miners. The competing crypto miner whoever solves these complex equations received the block reward, a fraction of digital Bitcoin. On the other hand, the PoS algorithm is validated in a deterministic fashion. Here, the more coins the crypto miner owns a virtual currency, the more the crypto miner is likely to be selected to validate blocks and add to the blockchain.

Dispersed estimation: The second important feature of blockchain is its dispersed estimation. Considering the example scenario of Bitcoin blockchain, each user of the Bitcoin blockchain that is executing a full node will possess an entire copy of the whole blockchain. The copy if composed of the data all the entire transactions recorded on the Bitcoin blockchain.

Upon successful completion of the download, the node are then said to be executed in an independent manner wherein the transactions are processed and are then sent through several nodes across the network. One of the most important aspects about the blockchain's dispersed estimation is that there is not central node in the network to process and distribute the data. However, every node forming the block in the blockchain is said to be executed in an independent manner.

Information repository: The third paramount feature of blockchain is the information repository. Here, the information repository refers to the information storage area. For example, in Bitcoin blockchain, the information repository here refers to the information related to the Bitcoin transactional data.

Source: Blockchains ensures source in a preprogrammed manner. In conventional type of banking, the customer is aware that the money deposited or the jewels kept in the locker are safe in the bank. However, in case of blockchain transaction, each activity performed by the user is traced, documented and fully identifiable without the aid of third-party required to attest for a specific action to be taken.

Stability: The blockchain transaction are said to be stable because of the reason that it is highly said to be stable. In that sense, the transaction pertaining to blockchain cannot be changed as soon as the transaction is recorded. Even in the case of erroneous record, the record cannot be deleted and is said to be noticeable cannot be deleted. Furthermore, in order to make corrections in the recorded transactions, a new transaction is said to be generated. That generated transaction is then said to refer the erroneous record.

1.1.2 BLOCKCHAIN AND BITCOIN TRANSACTIONS

Bitcoin on the other hand is a cryptocurrency. As far as Bitcoin is concerned, it is a form of digital currency in a decentralized form without the requirement of the third party or bank. Transactions are said to be performed between the users without the requirement of the third parties or the intermediaries. The verification process is performed via cryptography method and stored in a public distributed ledger called a blockchain.

1.1.3 HYPERLEDGER FRAMEWORKS

The hyperledger framework refers to an open source collective endeavor to provide cross industry blockchain technologies, hosted by Linux Foundation. Hyperledger simply refers to the type of software that everyone are said to use with the objective of creating one's own personalized blockchain. Figure 1.5 shows the figurative representation of hyperledger framework.

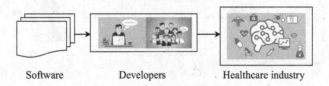

Software   Developers   Healthcare industry

FIGURE 1.5 Hyperledger framework for healthcare industry.

As shown in Figure 1.5, using personalized software with the support of specialized developers, software is developed for healthcare industry.

1.1.4 SMART CONTRACT FRAMEWORK AND ITS WORKING

The framework of smart contract is designed in such a manner that it allows to digitally permits, crosscheck, or validate the negotiation or contrast performance. The purpose of using smart contracts is that it allows the performance of credible transactions without the involvement of third parties or banks. The working of smart contract is illustrated with the aid of Figure 1.6.

As illustrated in Figure 1.6, the smart contract working is split into three steps and they are as follows:

1. Smart contracts are written in the form of code. With the code written, it is submitted to the blockchain. Both the code and the rules and regulations pertaining to the usage of the contract are available on the digital ledger in a public manner.
2. Whenever an event presented in the contract is triggered, to name a few are, expiry date or arrival of target price for purchasing a product, the event of code is said to be executed.
3. The regulators have the provision to look into the contract activity on the blockchain. This is performed with the objective of obtaining an understanding about the market. This is said to be attained with higher rate of privacy for the individual users.

Some of the uses of smart contract working are:

- Blockchain healthcare
- Blockchain music
- Blockchain government

Blockchain healthcare: With the specialization of smart contract, one of the fields of interest include healthcare, where the personal health records of the patients are said to be encoded and stored on the blockchain. Then, with the aid of only a private key, accessing of information is provided to specific individuals. Besides, the surgeries that has to be performed with certain patients are first stored on a blockchain and are then further sent

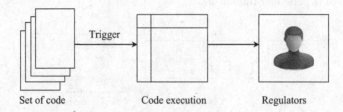

Set of code   Code execution   Regulators

FIGURE 1.6   Smart contract working.

to insurance providers as proof-of-delivery. The digital ledger besides are found to be useful for managing the healthcare of patients, specifically for drug supervision, maintaining rules and regulations, testing, and compiling with the healthcare supplies.

Blockchain music: Certain key issues in the case of music industry include the ownership rights, distribution of royalty in an equitable manner and ensuring transparency. Industry concerning digital music concentrates on production monetizing, however the rights pertaining to ownership are often ignored. The technology of blockchain and smart contracts technology cut shots this problem by forming an extensive and precise fragmented database of music rights. Besides, the digital ledger and clear transmission of artist royalties and real-time distributions to all involved with the labels are also being provided. In turn, the music players or vocalists are paid with digital currency based on the concerts being conducted by them or according to the terms of the contract.

Blockchain government: Digital data analysis say hackers garb the electronic system to influence votes. The digital ledger in blockchain technology would avert this at the votes are said to be in the form of encrypted nature. In this way, the private individuals can affirm that their valuable and sensitive votes were includes and guarantee who they elected for. In this way, by using the blockchain technology, the entire system would result in minimizing the money for the government too. Also, based on the report of McKinsey and Company made in 2013 report that open data that are freely attainable from freely accessible government-sourced data being available over the internet to all citizens of the country. Even startups have started using this blockchain technology to eliminate fraudulent schemes; farmers can utilize for precision farm-cropping; and so on.

1.2 BIG DATA SOURCE FOR BLOCKCHAIN

As discussed above in detail about the blockchain in the above section, to give a precise description, blockchain refers to a time-stamped series of data that is supervised by a group of computers and specifically not possessed by any single entity. As far as security is concerned, each block of data is said to be secured via cryptographic mechanisms. On the other hand, big data refers to the data sets that are bound to be complex for conventional type of data processing software [6–9]. Despite data with several rows ensure greater statistical power, however, data with higher complexity, i.e., involving more columns results in higher amount of false discovery rate. Figure 1.7 shows the conceptual diagram of Big Data and Blockchain.

As shown in Figure 1.7, the reason behind the successful relationship between Big Data and Blockchain is that the blockchain easily addressed the drawbacks of big data. Some of the reasons are:

- Decentralization
- Transparency
- Immutability

FIGURE 1.7　Conceptual diagram of big data and blockchain.

Decentralization: The data stored in a blockchain is not said to be owner by single entity. Hence, there is not probability of data or information getting lose if that entity is said to be compromised. The conventional way of data storage before the existence of Bitcoin was found to be centralized services. Here, the entity was found to be centralized that stored all the information pertaining to all the users in the network. Alternatively, while retrieving the information, this entity was said to be contacted whenever required with information, i.e., conventional banking system. In case of the decentralized system, the information is not stored at a single entity; on the other hand, every user in the network possesses his/her information. To perform communication with another user, it can be done directly, without the requirement of a third party or via another entity.

Transparency: The transparent framework of the blockchain assists in tracking the data back to its point of origination. Yet another interesting fact about the Big Data and Blockchain convergence is transparency. The person or the users' identity is said to be hidden by utilizing the complicated cryptography mechanism and described only via public address. Hence, while the users' real identity is secured, the transactions that were performed via the public address are only said to be viewed.

Immutability: Finally, immutability, in the framework of the blockchain refers to that once some transactions have been entered into the blockchain framework, the transactions are not said to be damaged or destroyed with. The reason behind immutability in blockchain is that of the application [10] of cryptographic hash functions. By applying this function, an input string of any length is obtained as input and producing an output of a definite length. As far as cryptocurrencies like Bitcoin are concerned, the transactions occurring through digital ledger are considered as input, processed via a hashing algorithm, resulting in a fixed length output.

1.2.1　Blockchain and Big Data to Secure Data

The technology combining Blockchain and Big Data assists in safeguarding the data from possible data leaks. With the information being stored on the channel, or blocks, even the most senior executives do not have the access to retrieve the information provided multiple permissions are obtained from the network to access the big data. Hence, it becomes highly complex and complicated from a cybercriminal point of view to seize it.

This is because of the reason that instead of uploading the big data or information to a database server or to a cloud server, blockchain subdivided into finite number of chunks and distributes them across the entire network of computers. In this way, the presence of middleman or third party is avoided for processing a transaction. Therefore, instead of placing the trust on a service provider, decentralization is said to be achieved or ensured via immutable ledger. Besides, the big data in blockchain is said to be in the encrypted form. Hence, it is said to be highly secured.

1.2.2 BLOCKHAIN AND BIG DATA TECHNOLOGIES FOR DATA ANALYSIS

Blockchain as referred to above as a digital and decentralized public ledger that records transactions across several machines linked in a peer-to-peer network basis. Though it was originally designed for cryptocurrency assets involving Bitcoin; however, in the current years it has been used for several purposes. Apart from the above said applications, blockchain also possess large amount of potentiality in analytics.

Several business establishments have started benefiting from data analytics for several years. One type of data analytics that is assured to revolutionize and metamorphose the industry is predictive analytics. Predictive analytics is concentrated on performing the predictions about the future course of actions on the basis of an immense amount of historical data as well as mechanisms using machine learning techniques.

Besides, with data analytics concern, the computational power of blockchain is possessed from collective associated computers. Hence it is said to be strong enough to precisely design the model to be examined on the basis of larger and enormous amount of datasets, stored across the computers and therefore the network and pull up the ones that can provide the answer. As for potential applications, blockchain analytics are found to be of heavily applicable in the area of marketing, i.e., digital marketing. In this way, even digital marketers could be able to prepare for future marketing advertisements with the assistance of data obtained from market realities.

1.2.3 BLOCKCHAIN FOR PRIVATE BIG DATA MANAGEMENT

Blockchain for private big data management is one of the main ways in which blockchain set itself from the conventional models of mechanisms that are frequently used today. No identity are said to be required in the network layer for blockchain. This means that not name, email id, address, or any other information pertaining to the user is required to download the information and the technology is started for utilization.

In other words it refers to that there requires or possess no third party of central server to store the users' information, making blockchain technology significantly more secure than a centralized server being used that can be easily cracked, putting its users' most sensitive data at risk. However, blockchain specifically increases the data analysis transparency. This is designed in such a manner that if an entry or

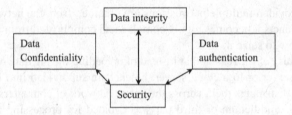

FIGURE 1.8 Conceptual diagram of security for blockchain.

transaction is not said to be verified, it is rejected in an automatic manner. Hence, the big data stored in blockchain is said to be highly and entirely transparent [11].

1.2.4 CONFIDENTIALITY, DATA INTEGRITY, AND AUTHENTICATION

Security concerns in blockchain for big data management are studied from the data confidentiality, data integrity and data authentication point of view. Figure 1.8 shows the conceptual diagram of security concerns in blockchain.

1.2.4.1 Data Confidentiality

This refers to the process of information hiding. It is performed with the purpose of hiding information in such a manner that only the intended recipients view the data or information provided by the provider. Data confidentiality is said to be attained by utilizing several data encryption mechanisms available in the industry and also through key pair models.

1.2.4.2 Data Integrity

This refers to the process of ensuring that the data transmitted from the source machine or user to the destination machine or user is invariable by all means. The data could be stopped in during the passage from the source machine or user to the destination machine or user and also can be altered. Nowadays, data integrity are said to be performed via by fingerprinting mechanism so that the destination machine or user could approve that the data or information is not changed.

1.2.4.3 Data Authentication

This refers to the process of ensuring that the data provider is obviously the provider of data. Therefore, the data provider should be authenticated in such a manner that the source machine or the sender bluffing to be the source machine would not be able to fake the communication.

1.2.4.4 Security Management Scenario for User Big Data in Blockchain

In this section a security management scenario for user big data in blockchain environment is provided. The scenario presented in Figure 1.9 combines several users controlled by Bigchain, the third parties inclined to access the information pertaining to private provided by blockchain users, around the blockchain infrastructure that is going to provide secure access to the data controlled by end users.

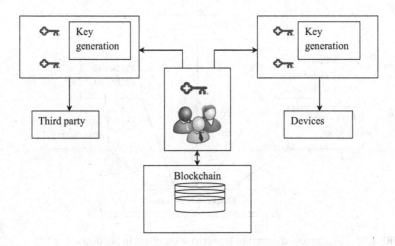

FIGURE 1.9 Security management scenario for block user data in blockchain.

As given in the above scenario, end users of blockchain environment administer their own security key. The security keys are generated using several cryptographic mechanisms to be provided to the blockchain users. Besides, generating security keys for blockchain users, credentials are also said to be generated for the third parties so that the data is said to be accessed under strict access control. Both credentials and security key are then translated to the blockchain environment so that rules defined by the end user with respect to storage of data and accessing of data is said to be ensured in a smooth manner [12–16].

The devices in blockchain in turn utilizing the security keys in turn store the information into the blockchain. Finally, the end users feed the blockchain with specific and relevant information, wherein the smart contract are then applied by applying restrictions based on the security factor corresponding to each blockchain user to store and retrieve data. Besides, the third party on the other hand, requesting authorization for the corresponding end user for utilizing private data can in turn access those data stored in the blockchain.

1.3 BLOCKCHAIN USE CASES IN BIG DATA

With increasing demand for data analytics, who can allocate more intuitions with data and assist in solving more complicated issues, the need of blockchain in data came into existence. This is even more complex and found to be more complicated when big data is concerned, a matured facet of data science that bestows with enormous data, cannot be lifted by conventional data processing models [17].

As far as blockchain and big data are concerned, data science examines data for analyses data for actionable perceptions, whereas blockchain documents and performs data validation. In other words, data science is used for prediction, whereas blockchain is used in data integrity. According to Maria Weinberger of Janexter, "if big is the quantity, blockchain is the quality." Therefore, blockchain is concentrated

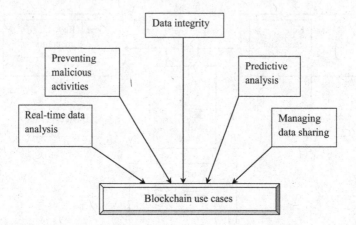

FIGURE 1.10   Conceptual diagram of blockchain use cases in big data.

on data validation whereas big data is concentrated on making predictions from enormous amounts of data. In this manner, blockchain has formed a new form of perspective in data management and operation, without relying on central angle where the entire data should be combined but in the form of decentralization.

In addition, validated data generated through blockchain technology comes in the form of structured manner and precise. Also, the other important aspect where the data generated through blockchain has become a boost for big data is the data integrity because blockchain verifies the source where the data came from via its linked chains without the aid of a third party (Figure 1.10).

Figure 1.10 shows the conceptual format of blockchain use cases in big data and involves:

- Ensuring data integrity
- Preventing malicious activities
- Predictive analysis
- Real-time data analysis
- Managing data sharing

The above said use cases in big data are described in the following sections.

1.3.1   Ensuring Data Integrity

The first aspect making the blockchain use in big data is that the data recorded on the blockchain are found to be reliable. This is because of the reason that they have undergone via a process involving verification, specifically meant for ensuring quality. Therefore, the blockchain is said to provide transparency also. Though blockchain technology was designed and processed with the objective of increasing the integrity of Bitcoin, with the popularity of Bitcoin, the underlying technology behind the design of blockchain has started receiving its popularity in other segment also. Figure 1.11 shows the data integrity for big data blockchain.

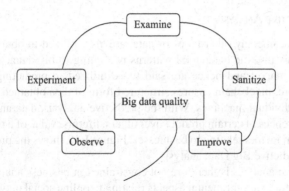

FIGURE 1.11 Big data blockchain data integrity.

Perhaps the most noteworthy evolution in the segment of IT sector over the past few years, the blockchain framework has the potentiality to transform the way the big data is conceptualized with higher level of both security and data quality. This is because the actions and transactions occurring on the blockhain framework are not said to be traced easily.

For example, Lenovo investigated this use case of blockchain framework to identify illegal documents and forms. The Lenovo company utilized blockchain framework for endorsing documents that were encoded using digital signatures. The digital signatures in turn were managed using the computers forming the network and the genuineness of the document was cross checked via a blockchain record. Blockchain, the technology that was initiated and designed with the objective of authenticating and tracking the Bitcoin transactions is said to be utilized for several purposes.

Several numbers of organizations and establishments have started utilizing blockchain for the purpose of authenticating their company data. The main driver for the application of blockchain is its security. In several cases, data integrity was said to be ensured when information pertaining to the originality of the transaction and interconnections concerning a data block were stored on the blockchain and validated in an automatic manner.

1.3.2 Preventing Malicious Activities

Because blockchain uses concurrence algorithm for authenticating the transactions, it is highly unsustainable for a single user to create a warning to the entire data network. A node or machine that starts to act anomalously are said to be straightforwardly recognized and removed from the network. This is because of the reason that the entire network is widely distributed in nature, making highly unmanageable for a single user to initiate ample computational power to change the acceptance factor and hence permitting unwanted data in the network. Therefore, to change the blockchain rules, large number of nodes or machines must be pooled together to create an agreement. This is not said to occur with a single user or machine.

1.3.3 PREDICTIVE ANALYSIS

Blockchain data, like any other types of data, are also utilized to obtain useful and valuable analysis into the behavioral patterns pertaining to big data, several trends existing in the market and hence are said to be utilized in predicting future outcomes. Therefore, blockchain ensures structured form of data obtained from several machines or individual machines. With the predictive analysis, data analyst obtains customer preferences of certain product over others, lifetime value of a product, price trends and churn factors relating to businesses. Figure 1.12 shows the process flow of Blockchain Predictive Big Data analysis.

The predictive analysis is therefore not constrained in business analysis, but also useful in predicting the sentimental aspects relating to online social networks, investment analysis, stock exchanges and so on. Besides, due to the distributed aspect of blockchain and the vast amount of computational capacity through it, with the presence of data scientists in smaller organizations can even conduct predictive analysis in an extensive manner. These data scientists connected on a blockchain framework utilizes the computational power of several thousand machines for analyzing the social outcomes in a wide scale.

The first application of predictive analytics through blockchain is said to be autologous. The predictive analysis are said to be of very useful in forecasting pricing trends for cryptocurrencies and other financial segments. This is because of the reason that predictive analysis is at the central of several marketing applications. To cite for example, customer segmentation used to take into consideration only preferences and rating coordinates in online social networks, while in the current scenario, new type of classifications can be structured via market actualities.

Besides, using predictive analytics, recommendation methods for systems for the hitherto specified classes also results in the increase of sale, customer preferences and satisfaction on a specific brand. By this way, the predictive analysis not only create a model to sell or purchase product but it is also said to regulate the price in a more progressive manner with the objective of keeping the price in a more comparative and competitive manner, therefore resulting in the increase in the profit margin.

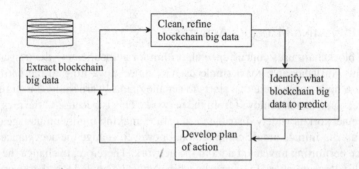

FIGURE 1.12   Blockchain predictive big data analysis.

1.3.4 REAL-TIME DATA ANALYSIS

As has been displayed in several financial and payment mechanisms, with the assistance of blockchain, real-time cross-border transactions have come into existence. Several banking firms and financial institutions have now started to examine blockchain. This is because of the reason that is provides swift usually, settlement of huge sum of money ensuring both affordability and real-time crossing the hurdles. In a similar manner, organization that necessitates real-time analysis of big data in a wide manner also utilizes blockchain-enabled system to achieve. With blockchain, even ecommerce sites and other business establishments started observing changes in big data in real-time analysis making it probable for fast decision making, ranging from blocking suspicious transaction to tracing out the involvement of abnormal events [18].

1.3.5 MANAGING DATA SHARING

With managing of data being shared, data obtained from research studies are said to be stored in a blockchain network. In this manner, software developers do not conduct data analysis that has been already conducted or wrongly repeat the data that has been previously used. However, the blockchain framework assists data scientists in monitoring the work, by analyzing the outcomes of trading on the blockchain environment.

Managing of data sharing in blockchain is composed of two types as shown in Figure 1.13. They are data block chain (DBC) and trading block chain (TBC). Here,

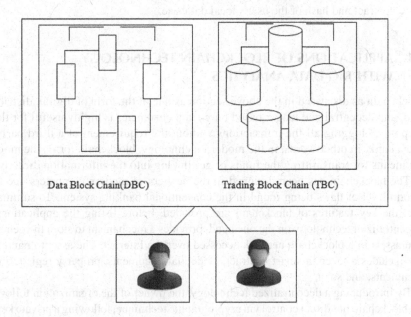

Data Block Chain(DBC)   Trading Block Chain (TBC)

FIGURE 1.13 Structure of managing data sharing.

the data block chain stores the actual user data and the trading block chain stores only the information that is of high regards for the purpose of trading and executing the transaction.

Each block as shown in the figure possesses a timestamp. The block in turn has hashed the previous block information. Besides, the transaction in each block are said to be verified. The blockchains, DBC and TBC is said to operate in an independent manner. Besides, both the blockchains exchange information pertaining to big data and authenticity is said to be attained via cryptographic techniques. The key to managing data sharing via a small case includes:

- Personal data sharing
- Smart contract
- Block creation

Personal data sharing: A controller shares a user's personal data via DBC and TBC. The controller also informs about this personal data sharing to the intended user. The processor involved in the data sharing mechanism again separates the personal information.

Smart contract: All the users in the network create a newsmart contract, responsible for holding terms, conditions, rules, and regulations for utilizing the personal information for data sharing. After all the processors or the user's agree upon the terms and conditions, processors or users are then permitted to access or retrieve the personal information via data sharing [19,20].

Block creation: Finally, a new block is created. The block includes a smart contract and hash of the user's local database.

1.4 APPLICATIONS OF BLOCKCHAIN TECHNOLOGY WITH BIG DATA ANALYTICS

Blockchain as discussed in the above section as one of the form of digital, distributed, and decentralized ledger based on virtual currencies is highly useful for the purpose of logging all the transactions without the requirement of a third party, like a bank. In other words, in the modern technology, blockchain forms the modern means for transmitting the funds or getting log into the information directly.

The blockchain has been designed in the new era of modern computers due to certain level of flaws being found in the conventional banking system. To summarize, the key features of blockchain are provided, before listing the applications. Decentralized technology on the one hand, provides a mechanism to store the generated assets in a blockchain network accessed over the Internet. The asset to name a few includes a token, a smart contract, evidential documents, property registration documents, and so on.

By introducing a decentralized technology, the owner of the organization following blockchain has direct control via cryptographic technique, following a private key, directly correlated to the asset. The owner of the organization with the decentralized technology followed with blockchain transfer the asset whenever desired and to any-one. To mention in specific, the reasons for decentralization include:

- Authorized users
- Fault tolerance
- Scam-free mechanism
- Third party removal
- Lower transaction cost and time

Next, blockchain being a public digital ledge provides mechanism for giving the valuable and most sensitive information of all users and all the digital transactions involved by the users that have ever been executed. This blockchain technology in turn assists in documenting each transaction shared across the network throughout the globe. With this, every user in the organization has the potential to validate the transactions and maintain an identical digital ledger copy. Some of the attractive benefits include:

- Privacy preservation
- Stopping of fraudulent activities
- Removal of mediator
- Cut costing speed

Next, an important aspect of blockchain remains in the authentication provided because several institutions possess sensitive documents, and contracts have to be safeguarded against malicious activities or organizations. To safeguard the most sen-sitive documents, a hash is said to be created in blockchain for each file, unique for every single transaction.

- Header part
- Transaction part

Specifically, the community modern era is looking the banking institutions as third parties and stealing the fees in the name of transaction. However, in case of blockchain, real-time transactions presumably reduce transaction fees (as shown in Figure 1.14).

Blockchain technology ensures absolute engine to provide a novel concept for the new connected world in the Internet era. While digital ledgers and identities on the one hand provide an immense and inevitable means of the connected world, the method and mechanism through white online information in blockchain technology is secured is under intense scrutiny. As illustrated in Figure 1.14, some of the applica-tions of blockchain technology with big data analytics include:

- Anti money laundering
- Cyber security
- Supply chain monitoring
- Financial AI systems
- Medical records

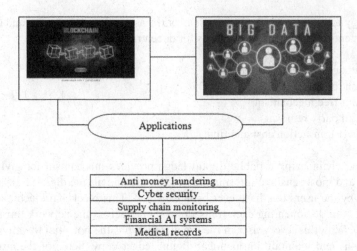

FIGURE 1.14   Applications of blockchain with big data analytics.

1.4.1 ANTI MONEY LAUNDERING

Blockchain mechanism and its digital ledger permits for higher amount of transparency with smart contractors or regulators revamping the reporting process. In addition, the distributed and immutable digital ledge permits in changeless transaction history. Besides, the digital ledger also acts as a pivotal hub for storing the big data with the purpose of processing large transactions. By performing anti money laundering, it can act with the undertaking beyond with the risk officers within the financial institutions.

Cryptocurrencies ensure the user around the world with immediate, secure, and money where the blockchains provide both static and dynamic storage of record for their corresponding transactions. Conventional system only required the use of central authority wherein the monetary supply and payment transfer were said to be performed without the tampering possibilities. On the other hand, blockchain mechanisms prevent this method of payment transfer by providing third party environment for ensuring payment transfers, thus forming a peer to peer environment.

To achieve anti money laundering, nowadays in every business establishments, refined identity management utilizing encryption-based mechanism is being structured in a decentralized manner. In addition, identity management based on digital signature also assists the banking and financial sectors meet the requirements of the ever-changing know your customer (KYC) and customer due diligence (CDD) requirements. With this, the costs incurred in robust KYC are also found to be minimized and finally, crimes related to finance and violations related to consent could be reduced in the long run.

Smart contracts in the current years with the use of blockchain for big data have started metamorphosing the conventional lending system. For example, traditional money lenders service the borrowers with the required loans, while instantly imposing higher rate of interest for the borrowed loan amount and attesting their property in the form of collateral. Due to this, larger number of borrowers have fall into bankruptcy and also started losing their home or property. With the evolution and introduction of blockchain technology, this is undercut by eliminating a stranger to loan the required money or property as collateral. In this way, there requires not need of showing the credit history by introducing blockchain. No need to show the lender credit or work history and also manual processing of several documents has been also cut shorted.

1.4.2 CYBER SECURITY

The blockchain mechanism in the current era is found or said to be occupied in a wider manner in every walks of our lives, ranging from banking to healthcare and beyond. Besides, this, the cybersecurity is an industry that has received wider attention with the introduction of this blockchain technology [10] with an extent for more in the recent future. Hence, by eliminating much of the human effort put from the angle of data storage, blockchain technology extensively avoid the risk of error involving humans that forms the main cause of data theft. The reason behind the success of this blockchain technology is any form of digital transactions are said to be put into the blockchain and the specification of the industry does not remaining the main concern. Also, by introducing blockchain technology in any types of industry can eliminate the risk involving data theft to a larger extent; therefore, the nature and size of data remains private and secure.

1.4.3 SUPPLY CHAIN MONITORING

Blockchain also comes in specifically easy-to-use when it appears to monitoring the supply chains involved in the organization. By eliminating paper-based trails, institutions and organization should be in a position to point out the inefficiencies within their supply chains [21] in a swift manner and also to discover items in real time. Besides, blockchain technology would permit institutions and organizations to watch how products or services executed from the angle of quality control from the place of origin to destination.

The probabilities for application of the blockchain technology in big-data supply chain solutions are that it monitors goods status as they are in transportation. Here, the data pertaining to all the customer or user is found to be available with all the users in real time. Some of the benefits for supply chain monitoring [22] remains in the verification of product labeling claims and that of product origins. Also, one of the most important and paramount remains in the probability of providing human rights along with fair wages.

For example, blockchain permits banks, financial institutions, and companies in a longer run have started end-to-end visibility of their supply chain by means of providing larger amount of data in the range of millions or trillions pertaining to several location and condition of the supplies by transporting them across the globe. According to the status of 2016, a Deloitte and MHI [23] report surveyed 99 leading supply chain companies and identified that the sensors were utilized nearly by 44% of these respondents. Among them, nearly 87% of these institutions said they are supposed to use this technology by 2020. Besides, the blockchain technology also stores, supervises, safeguards, and transfers this smart information.

1.4.4 FINANCIAL AI SYSTEMS

As far as financial transactions are concerned, blockchain technology with big data is set to become an extensive aspect involving monetary transactions. Besides, financial transactions, big data, and blockchain are found to be in synchronous with each other in supplying the products and services related to the financial services [24] industry. This is because of the reason that the conventional form of financial systems are found to be both complex in terms of time and cost involved, therefore resulting in erroneous information. Besides, intermediaries or the third party are also required to conduct or perform the process so that in case of conflicts, it can be resolved in a smooth manner. This in turn results in higher amount of cost, stress, time stamp, money depending on the nature and size of data involved. Due to this several banking and financial sectors have started using this blockchain technology with the purpose of introducing innovations like, smart contracts, smart bonds, and so on.

1.4.5 MEDICAL RECORDS

Medical records [25] are found to be an area where records are found to be highly critical and therefore have to be inspected in a continuous manner. When the big-data systems that enter into this data-oriented sector is designed with the aid of blockchain technology, all records are then said to be concealed with a clear track record and also maintained with higher rate of transparency.

REFERENCES

1. A. Stavrou, R. Ramadoss, J. Rupe, C. Rong, T. Kostyk, and S. Chandrasekaran, IEEE blockchain- future directions initiative, IEEE Access, July 2018.
2. T. Rana, A. Shankar, M. K. Sultan, R. Patan, and B. Balusamy, An Intelligent Approach for UAV and Drone Privacy Security Using Blockchain Methodology. In *2019 9th International Conference on Cloud Computing, Data Science & Engineering (Confluence)*, (pp. 162–167). IEEE, January 2019.
3. S. Kim, G. Chandra Deka, P. Zhang, *Role of Blockchain Technology in IoT Applications*, Vol. 115. 1st ed., Academic Press, Cambridge, MA, 2019.
4. M. M. Hassan Onik, C.-S. Kim, N.-Y. Lee, and J. Yang, *Privacy-Aware Blockchain for Personal Datasharing and Tracking*, Springer, Germany 2019.
5. M. Andonia, V. Robua, D. Flynna, S. Abramb, D. Geachc, D. Jenkinsd, P. McCallumd, and A. Peacock, Blockchain technology in the energy sector: A systematic review of challenges and opportunities, *Renewable and Sustainable Energy Reviews*, 2019, 100, 143–174.
6. P. Rizwan, M. R. Babu, B. Balamurugan, and K. Suresh, Real-time big data computing for internet of things and cyber physical system aided medical devices for better healthcare. In *2018 Majan International Conference (MIC)* (pp. 1–8). IEEE, March 2018.
7. R. Patan, and M. R. Babu, A novel performance aware real-time data handling for big data platforms on Lambda architecture. *International Journal of Computer Aided Engineering and Technology*, 2018, 10(4), 418–430.
8. R. Patan, and S. Kallam, Performance improvement IoT applications through multimedia analytics using big data stream computing platforms. In *Exploring the Convergence of Big Data and the Internet of Things* (pp. 200–221), 2018, IGI Global, Hershey, PA.

9. P. Rizwan, and M. R. Babu, Performance improvement of data analysis of IoT applications using re-storm in big data stream computing platform. *International Journal of Engineering Research in Africa*, 2016, 22, 141–151. Trans Tech Publications.

10. F. Casino, T. K. Dasaklis, and C. Patsakis, A systematic literature review of blockchain-based applications: Current status, classification and open issues, *Telematics and Informatics*, 2019, 36, 55–81.

11. G. Nagasubramanian, R. K. Sakthivel, R. Patan, A. H. Gandomi, M. Sankayya, and B. Balusamy, Securing e-health records using keyless signature infrastructure blockchain technology in the cloud. *Neural Computing and Applications*, 2018, 32(3), 639–647.

12. M. Khari, A. K. Garg, A. H. Gandomi, R. Gupta, R. Patan, and B. Balusamy, Securing Data in Internet of Things (IoT) Using Cryptography and Steganography Techniques. *IEEE Transactions on Systems, Man, and Cybernetics: Systems*, 2019.

13. S. Karthikeyan, R. Patan, and B. Balamurugan, Enhancement of security in the Internet of Things (IoT) by using X. 509 Authentication Mechanism. In *Recent Trends in Communication, Computing, and Electronics* (pp. 217–225), 2019, Springer, Singapore.

14. A. Shankar, N. Jaisankar, M. S. Khan, R. Patan, and B. Balamurugan, Hybrid model for security-aware cluster head selection in wireless sensor networks. *IET Wireless Sensor Systems*, 2018, 9(2), 68–76.

15. S. Karthikeyan, P. Rizwan, and B. Balamurugan, Taxonomy of security attacks in dna computing. In *Advances of DNA Computing in Cryptography* (pp. 118–135), 2018, Chapman and Hall/CRC, Milton, FL.

16. S. Namasudra, D. Devi, S. Choudhary, R. Patan, and S. Kallam, Security, privacy, trust, and anonymity. In *Advances of DNA Computing in Cryptography* (pp. 138–150), 2018, Chapman and Hall/CRC, Milton, FL.

17. S. R. Kumar, N. Gayathri, S. Muthuramalingam, B. Balamurugan, C. Ramesh, and M. K. Nallakaruppan, Medical big data mining and processing in e-Healthcare. In *Internet of Things in Biomedical Engineering* (pp. 323–339), 2019. Academic Press, San Diego, CA.

18. T. Poongodi, M. S. Khan, R. Patan, A. H. Gandomi, and B. Balusamy, Robust Defense scheme against selective drop attack in wireless ad hoc networks. *IEEE Access*, 2019, 7, 18409–18419.

19. R. Krishnamurthi, R. Patan, and A. H. Gandomi, Assistive pointer device for limb impaired people: A novel Frontier Point Method for hand movement recognition. *Future Generation Computer Systems*, 2019, 98, 650–659.

20. A. Selvaraj, R. Patan, A. H. Gandomi, G. G. Deverajan, and M. Pushparaj, Optimal virtual machine selection for anomaly detection using a swarm intelligence approach. *Applied Soft Computing*, 2019, 84, 105686.

21. H. Min, Blockchain technology for enhancing supply chain resilience, *Business Horizon*, Elsevier, the Netherlands, 2019.

22. W. Gao, W. G. Hatcher, and W. Yu, A survey of blockchain: Techniques, applications, and challenges, *27th International Conference on Computer Communication and Networks (ICCCN)*, IEEE, Hangzhou, China, 2018.

23. T. Robles, B. Bordel, R. Alcarria, and D. Sánchez-de-Rivera, Blockchain technologies for private data management in AmI environments, *Proceedings*, 2018, 2, 1230. doi:10.3390/proceedings2191230

24. A. P. Joshi, M. Han, and Y. Wang, A survey on security and privacy issues of blockchain technology, *American Institute of Mathematical Sciences*, 2019, 1(2), 121–147.

25. A. Ali Siyal, A. Zahid Junejo, M. Zawish, K. Ahmed, A. Khalil, and G. Soursou, Applications of blockchain technology in medicine and healthcare: Challenges and future perspectives, *Cryptography*, 2019, 3(1), 3.

2 Blockchaining and Machine Learning

R. Venkatesh, L. Godlin Atlas,
and C. Mageshkumar

CONTENTS

2.1 BLOCKCHAIN FOREWORD

In the recent past, everyone everywhere went crazy about cryptocurrencies and in particular, the so-called cryptogold—Bitcoin. However, even though that your (grand) parents have probably have heard about this digital currency, it's likely that they haven't got a single clue about the fascinating technology behind it called blockchain.

But what it is that makes blockchain so special that more and more businesses in various industries are now adopting it? That's what I will try to explain you.

When we think about technology, we assume the constant buzzing notifications from your smartphone and people walking on the streets, while staring at a smartphone screen. However, blockchain is the first technology, which will actually connect us instead of the opposite. It will change the way we make decisions and exchange value [1].

Our ancestors have traded by using violence and social repercussions. With time, society evolved and government institutions and other intermediaries took over the way we trade and exchange value. But in order to understand one of the many ways that blockchain is of advantage to our society, lets take a look at the banking system. If you want to make a transaction, the bank is the institution which sets the rules on how this transaction will be executed—how long will it take, how much will they charge you and so on. If this transaction is made on the blockchain, there are a set of computers which are participating in the network to validate this transaction. Figure 2.1 shows the transaction one-to-one without the bank as an intermediary.

This fancy term blockchain actually means a block of data that has been recorded over a certain amount of time and is grouped and cryptographically linked to a previous set of data forming a chain of events. These computers are agreeing upon what

Like a spreadsheet in the sky

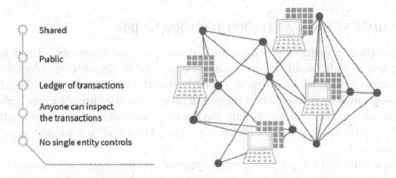

Shared

Public

Ledger of transactions

Anyone can inspect
the transactions

No single entity controls

FIGURE 2.1 Transaction one-to-one without the bank as an intermediary. (From S. Voshmgir, Token Economy, 2019, https://blockchainhub.net.)

happened over a time period and then each of them represents that data instead of having one centralized entity, that is doing so. All of these events which occurred on the blockchain are recorded on a public ledger.

If and when all parties to the smart contract fulfill the predefined arbitrary rules, the smart contract will auto execute the transaction. These smart contracts aim to provide transaction security superior to traditional contract law and reduce transaction costs of coordination and enforcement.

Smart contracts can be used for simple economic transactions like sending money from A to B. They can also be used for registering any kind of ownership and property rights like land registries and intellectual property, or managing smart access control for the sharing economy, just to name a few. Furthermore, smart contracts can be used for more complex transactions like governing a group of people that share the same interests and goals. Decentralized autonomous organizations (DAOs) are such an example for more complex smart contracts.

With blockchains and smart contracts, we can now imagine a world in which contracts are embedded in digital code and stored in transparent, shared databases, where they are protected from deletion, tampering, and revision. In this world, every agreement, every process, task, and payment would have a digital record and signature that could be identified, validated, stored, and shared.

Intermediaries like lawyers, brokers, and bankers, and public administrators might no longer be necessary. Individuals, organizations, machines, and algorithms would freely transact and interact with one another with little friction and a fraction of current transaction costs.

Therefore, blockchains and smart contracts:

- Radically reduce transaction costs (bureaucracy) through machine consensus and auto-enforceable code.
- Bypass the traditional principal-agent dilemmas of organizations, thus providing an operating system for what some refer to as "trustless trust."

This means that you don't have to trust people and organizations, you trust code, which is open source and provides transparent processes.

2.2 BLOCKCHAIN AND CRYPTOCURRENCIES

Blockchain is a distributed ledger that is encrypted and immutable. Each new block that is added to the chain needs to be verified by the previous block with a unique identifier. Blockchains are cloud-hosted (hence the term "distributed"). So, while this is obviously useful for transactions and has traditional finance institutes scrambling because blockchain can cut the middle man, blockchain's native safety features make it a great choice for what's known as "the single source of truth" as well.

Blockchain technology and cryptopayments are the future of e-commerce. Not only for big web shops but for everyone. Blockchain will solve a lot of e-commerce problems. Such as the following:

- **Smart and honest reviews**: Blockchain assures the transparency in terms of honest shopping. Shoppers will be able to leave a review only after receiving ordered goods, it means no fake feedbacks.
- **Affiliate marketing on blockchain**: Sellers will be able to cooperate with bloggers and get a free promotion from them. Traffic owners (bloggers) will be able to monetize their reviews through an affiliate program.
- **Fast and cheap transactions**: Traditional online global trading implies long and costly transactions. Multicurrency wallet will help shoppers and vendors to interact easily and perform cross-border payments in a convenient way. A fee for transferring cryptocurrency is much lower than for transferring fiat money.
- **Safe shopping**: Buyers will be safeguarded due to usage of smart contract system. Smart contract transfers money to a seller's account only when an ordered item has reached its destination. So, a customer pays only after receiving the product, it increases his or her confidence in a service and makes online shopping more attractive.

Moreover, blockchain brings cryptocurrency to common usage in online trading. It means that selling globally will be easier because it's more convenient for people.

2.2.1 THE DIFFERENT TYPES OF BLOCKCHAINS

There are three primary types of blockchains, which do not include traditional databases or distributed ledger technology (DLT) that are often confused with blockchains.

1. Public blockchains like Bitcoin and Ethereum
2. Private blockchains like Hyperledger and R3 Corda
3. Hybrid blockchains like Dragonchain (Figure 2.2)

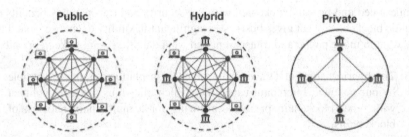

FIGURE 2.2   Different types of blockchain.

2.2.1.1   What Is a Public Blockchain?

Let's explore the different types of chains. And start with public blockchains, which are open source. They allow anyone to participate as users, miners, developers, or community members. All transactions that take place on public blockchains are fully transparent, meaning that anyone can examine the transaction details.

1. Public blockchains are designed to be fully decentralized, with no one individual or entity controlling which transactions are recorded in the blockchain or the order in which they are processed.
2. Public blockchains can be highly censorship-resistant, since anyone is open to join the network, regardless of location, nationality, etc. This makes it extremely hard for authorities to shut them down.
3. Lastly, public blockchains all have a token associated with them that is typically designed to incentivize and reward participants in the network.

2.2.1.2   What Is a Private Blockchain?

Another type of chains are private blockchains, also known as permissioned blockchains, possess a number of notable differences from public blockchains.

1. Participants need consent to join the networks.
2. Transactions are private and are only available to ecosystem participants that have been given permission to join the network.
3. Private blockchains are more centralized than public blockchains.

Private blockchains are valuable for enterprises who want to collaborate and share data, but don't want their sensitive business data visible on a public blockchain. These chains, by their nature, are more centralized; the entities running the chain have significant control over participants and governance structures. Private blockchains may or may not have a token involved with the chain.

2.2.1.3   What Is a Hybrid Blockchain?

Dragonchain occupies a unique place within the blockchain ecosystem in that it's a hybrid blockchain. This means that it combines the privacy benefits of a permissioned and private blockchain with the security and transparency benefits of a public blockchain. That gives businesses significant flexibility to choose what data they want to make public and transparent and what data they want to keep private.

1. The hybrid nature of Dragonchain blockchain platform is made possible by our patented Interchain capability, which allows us to easily connect with other blockchain protocols allowing for a multi-chain network of blockchains.
2. This functionality makes it simple for businesses to operate with the transparency they are looking for, without having to sacrifice security and privacy.
3. Also, being able to post to multiple public blockchains at once increases the security of transactions, as they benefit from the combined hash power being applied to the public chains [2].

2.3   WHY DO WE NEED BLOCKCHAIN TECHNOLOGY?

Now that we understand what a blockchain is and the different types of blockchains let's discuss why we even need blockchains to begin with. There are a variety of blockchain use cases and benefits to blockchain implementation, the most well-known being value transfer over the Bitcoin protocol. For cryptocurrencies like Bitcoin, blockchain solves a very specific problem that had hampered previous efforts at developing a digital currency. That problem is known as the double spend phenomenon. We all understand that the typical way in which we share things in the digital world is to create a copy of what we have, such as a pdf or image, and sending that to another person.

1. As you can imagine, if this pdf were a dollar, both the sender and recipient would have identical copies of this dollar and conceivably could both spend it.
2. Blockchain technology solved this by ensuring the recipient knows that only they have the dollar and the sender knows that they no longer have it.
3. Anyone who tries to spend the dollar knows that only the next recipient now has the dollar.

2.3.1   Why is Blockchain Technology Unique?

Any business transaction, including information transfer, requires trust and reciprocity, and fast and effective exchanges—including big data exchange. Currently, centralized authorities (e.g., regulators and auditors, government authorities, and banks) provide a backbone of integrity [3,4].

Blockchain technology provides a mechanism to add more trust, providing transparency and reducing the need for so many intermediaries. Blockchain technologies shine with these unique strengths:

- **Flexibility**: Blockchains store any kind of digital information, including computer codes that can be executed after the appropriate parties enter their keys.

- **Decentralized and distributed structure**: "No single individual or company has control of the data entry or its integrity. The accuracy of the blockchain is continuously verified by each computer in the network."
- **Enhanced security**: Blockchain is very hard to hack. Should a record's data change, this could be retroactively examined, revealing answers about when and who.

These blockchain advantages translate to streamlined business processes, new business models, and industry transformation way beyond bitcoins and into the realm of data asset management.

2.4   BLOCKCHAIN: A CURRENCY FOR DATA ASSET EXCHANGE

Blockchain will become a prominent currency for data asset exchange in 2019. In his presentation on "The Blockchain Scenario," David Furlonger, predicted that blockchain solutions will be "designed to address a specific operational issue—most often in terms of inter-organizational process or record keeping inefficiency." This has special relevance in data transfer.

For example, information trade among distributers, partners, and customers within the supply chain and using blockchain are proving successful. McKesson, a leading healthcare company for wholesale medical supplies and equipment, needed to trace certain prescription drugs to comply with the Drug Supply Chain Security Act (DSCSA). The company is working with the industry toward the ability to verify returns for the November 2020 deadline and is looking toward the November 2023 full serialization requirements [5].

Scott Mooney, vice president, distribution operations for McKesson Pharmaceuticals, sees blockchain as an "aid in the lookup directory for returns" and to "perhaps aid in facilitating gathering of DSCSA data from off-block data repositories." Other business models, like supply chains, promise to be just as useful.

The databases responsible for content management transfer and licensing intellectual property will start to make more use of blockchain technologies in 2019. Microsoft and Ernst & Young (EY) created an "intellectual property blockchain enabling companies and individuals to clearly specify, account for, and track the attribution of digital content throughout the network of stakeholders involved in the development and release of a video game" [6].

Blockchain can seal agreements between parties through "smart contract" and this use is becoming more technologically feasible. Successful emerging use cases in the creative economy will take off in 2019, allowing for better data management of these agreements.

2.4.1   DIGITIZATION OF GOVERNMENT SERVICES

Blockchain will drive government services toward digitization in 2019. Blockchain could be used in drawing up birth certificates, passports, and similar documents and making them easily available to citizens. Already the Illinois Blockchain Initiative has created a pilot database with birth certificate and identity information to be used in government services. Australian citizens use blockchain as a consumer and provider to a local power grid. New York City enables authoritative, trusted, and immutable social service transactions [3].

Gartner predicts identities, voting, public records, and citizen transactions will become transformed by blockchain. If executed well, blockchain would reduce the call for voting recounts and supervising auditors.

"By casting a vote as a transaction, blockchain can be used to track the votes. By using blockchain, the integrity of the vote count can be trusted. Because blockchain leaves an audit trail, it can be verified no votes were removed or altered, and no improper votes were added."

Switzerland conducted a successful test vote and plans on leading the way, with e-voting in at least 26 cantons by October 2019. In fact, Europe may be the pioneers. In a press release describing a Blockchain Partnership among European countries, Mariya Gabriel, Commissioner for Digital Economy and Society, says:

> In the future, all public services will use blockchain technology. Blockchain is a great opportunity for Europe and Member States to rethink their information systems, to promote user trust and the protection of personal data, to help create new business opportunities and to establish new areas of leadership, benefiting citizens, public services and companies.
>
> It's likely more cities, states, and regions, especially in Europe, will use blockchain.

2.4.2 Growth of Data Banks Using Blockchain

Consumer opinion has shifted. They trust brands but want more control over their data. Blockchain offers promise here. Consumers with access to the ledger can see transactions, like e-accounts, and manipulate their data through their blockchain key.

Any blockchain-based application needs to live up to protecting consumers and providing the right information to the right people at the right time. That is a tall and complex order for businesses. Organizations could emerge to act as an intermediary, handling the complexities of data asset management, reducing security breaches, and complying with laws like the GDPR.

Imagine if consumers could choose from a buffet of companies and services responsible for securing and ensuring their data goes to the right people. For example, medical information would not exist at a specific clinic or be centralized by doctors and pharmacies. A third-party company would house that type of big data. The consumer would inform this data bank which businesses can have access to specific types of health data. They could walk into any clinic and provide a blockchain key, and a medical office could download the information for a set time, after which the data is locked. This type of business model could gain traction in 2019.

Already TRON, Caspian, MediBloc, and other startups have created decentralized marketplaces where people can store and consume content without a centralized company, say Netflix. These types of services and related business services will be something to watch through 2019.

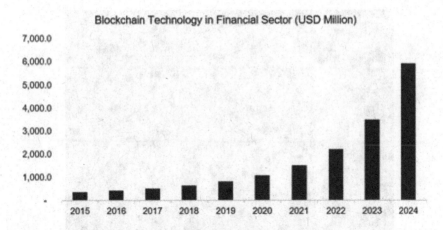

FIGURE 2.3 Blockchain growth in financial sector.

In 2019, blockchain technologies will mature. It will become the currency of information transfer and increasingly digitize governments. Businesses opportunities will grow that allow consumers manage their data while relieving a wide range of businesses from very detailed and complex data management and governance administration. The structure around blockchain will make this possible.

Alison DeNisco Rayome writes that "a lack of standards and interoperability between different platforms and solutions, legal and regulatory concerns around data privacy and intellectual property, and the technical complexity" barricade blockchain technologies from mainstream use. She notes, citing the Deloitte report, that there have been signs of progress in these areas from some regulatory bodies and consortia.

Blockchain requirements will have to solidify, because of a push from governments and the financial sector. For example, blockchain technologies, in finance, is anticipated "grow at a CAGR of 38.6 percent" over the next six years (Figure 2.3).

These types of standards will shape the blockchain's role in big data, government digitization, and business opportunities that increasingly use decentralized industries.

2.5 FOUR TRENDS THAT BLOCKCHAIN WILL DOMINATE IN EMERGING TECH IN 2019

2.5.1 TRADITIONAL COMPANIES LOOKING INTO BLOCKCHAIN TECHNOLOGY

We see more and more companies using this cutting-edge technology or are in the process of implementing it given its myriad applications. Organizations are now building capabilities that are needed to push blockchain into mainstream adoption. Industries that are seeing an increased adoption of blockchain are banking, financial services, insurance, supply chain management, healthcare, e-commerce, gaming, and academics.

FIGURE 2.4   Traditional companies looking into blockchain technology.

The Society for Worldwide Interbank Financial Telecommunication (SWIFT) has launched a pilot global payment initiative (GPI) service to join the growing blockchain and fintech services (Figure 2.4).

Cost reduction and process simplification are strategic values that blockchain brings to the table. We've seen the adoption of blockchain in payments, remittances, provenance, and traceability in the early days which is consistent with the expected returns. While there are applications of blockchain that promise to deliver topline advantage to firms, in 2018 and 2019, leaders are expected to continue adoption of cases with clear bottom-line benefits.

2.5.2   REAL-WORLD USE CASES OF BLOCKCHAIN TECHNOLOGY BEYOND FINANCIAL TRANSACTIONS

We are also now seeing some real use cases or proof of concepts being rolled out in other industries. To name a few topline benefits, blockchain helps prevents perjury in supply chain management as it allows the record of transactions to be maintained in the form of an immutable ledger. Picking on it various advantages, an Australian truck and transport insurance company National Transport Insurance (NTI) has launched a pilot program to deploy blockchain technology for end-to-end tracking of Australian beef exports (Figure 2.5).

Healthcare is seeing a rising traction of blockchain adoption as it enables creating records of patients' treatments and provide doctors with relevant information by bringing the entire information online in a much secured manner. Blockchain has found inroads in talent management as well. The SP Jain School of Global Management, one of India's top-ten business schools, has issued 1,189 blockchain-based certificates to graduates who recently obtained degrees and professional certifications. The certificates which are now live on the Ethereum blockchain will

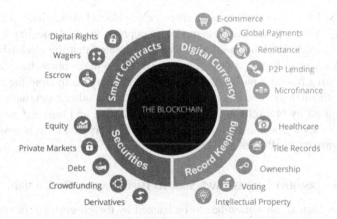

FIGURE 2.5   Real-world use cases of blockchain technology.

allow prospective employers and other parties to verify the authenticity of a job seeker's educational qualifications without having to contact the business school.

Blockchain enables secured transactions as it is encrypted and hence it is a great alternative in areas such as identity management. With its array of benefits, it is safe to say that it will be instrumental in trade, fraud detection, policy management in insurance, government and public sector as well as food safety and provenance in agriculture.

2.5.3   INNOVATIVE STARTUPS WILL DISRUPT THE INDUSTRY WITH BLOCKCHAIN TECH

Fast growing startups have found themselves in the position to be able to disrupt other businesses in their sector. They are headed by executives who are experienced and well-connected in their industries (Figure 2.6).

FIGURE 2.6   Innovative startup.

These are the businesses that are often able to apply blockchain technologies in ways that are truly a part of their business model, as opposed to supplementing it.

A 2018 blockchain global survey by Deloitte revealed that the established companies face a host of legacy concerns and are trying to make blockchain fit into an already existing business paradigm that may or may not benefit from the introduction of this technology. The emerging disruptors, on the other hand, have business models inspired by blockchain. They are experimenting and building without the constraints of legacy business processes. They focus energy on what is possible and then deal with any challenges as they rise.

2.5.4 Government Bodies Have Started Implementing Blockchain

Every technological innovation has to be backed by the government in some form or another and blockchain is no exception! In order to bring blockchain to a wider audience, it has to be legalized by the government, more so because its potential applicants such as banking, insurance, pharma, etc., operate within the boundaries of the current regulatory norms.

A number of initiatives have been taken across jurisdictions both at federal and state/provincial levels to implement blockchain to improvise on services and transform inter-governmental and citizen transactions. Areas like titles transfer and identification have gained maximum traction for blockchain in the government sector (Figure 2.7).

In geographies like Switzerland, Thailand, and Canada, the government has successfully deployed blockchain to improvise processes. To name a few, Swiss National Postal Service and Telecom Leader are building a blockchain platform for their own blockchain-based applications, Thailand revenue department are tracking VAT payments using blockchain. Very recently, Dubai's official government credit bureau announced that they are introducing digital payments to the public sector.

FIGURE 2.7   Government bodies have started implementing blockchain.

While a lot of blockchain initiatives have seen the light of day many are still in the pilot stages. In 2020, these initiatives will pick up pace sooner than we know!

2.6   MACHINE LEARNING INTRODUCTION

The term machine learning (ML) was coined by Arthur Samuel in 1959, an American pioneer in the field of computer gaming and artificial intelligence and stated that "it gives computers the ability to learn without being explicitly programmed."

And in 1997, Tom Mitchell gave a well-posed mathematical and relational definition that "A computer program is said to learn from experience E with respect to some task T and some performance measure P, if its performance on T, as measured by P, improves with experience E."

Machine learning is a latest buzzword floating around. It deserves to, as it is one of the most interesting subfield of computer science. So what does machine learning really mean?

Let's try to understand machine learning in layman terms. Consider you are trying to toss a paper to a dustbin.

After first attempt, you realize that you have put too much force in it. After the second attempt, you realize you are closer to target but you need to increase your throw angle. What is happening here is basically after every throw we are learning something and improving the end result. We are programmed to learn from our experience.

This implies that the tasks in which machine learning is concerned offers a fundamentally operational definition rather than defining the field in cognitive terms. This follows Alan Turing's proposal in his paper "Computing Machinery and Intelligence," in which the question "can machines think?" is replaced with the question "can machines do what we (as thinking entities) can do?"

Within the field of data analytics, machine learning is used to devise complex models and algorithms that lend themselves to prediction; in commercial use, this is known as predictive analytics. These analytical models allow researchers, data scientists, engineers, and analysts to "produce reliable, repeatable decisions and results" and uncover "hidden insights" through learning from historical relationships and trends in the data set (input).

Suppose that you decide to check out that offer for a vacation. You browse through the travel agency website and search for a hotel. When you look at a specific hotel, just below the hotel description there is a section titled "You might also like these hotels." This is a common use case of machine learning called "recommendation engine." Again, many data points were used to train a model in order to predict what will be the best hotels to show you under that section, based on a lot of information they already know about you (Figure 2.8).

So if you want your program to predict, for example, traffic patterns at a busy intersection (task T), you can run it through a machine learning algorithm with data about past traffic patterns (experience E) and, if it has successfully "learned," it will then do better at predicting future traffic patterns (performance measure P).

The highly complex nature of many real-world problems, though, often means that inventing specialized algorithms that will solve them perfectly every time is

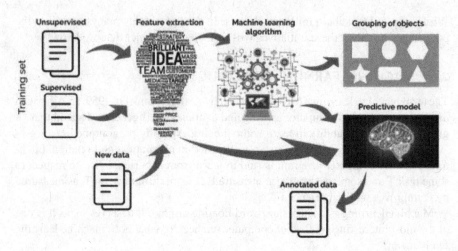

FIGURE 2.8 Types of machine learning algorithm.

impractical, if not impossible. Examples of machine learning problems include, "Is this cancer?", "Which of these people are good friends with each other?", "Will this person like this movie?" such problems are excellent targets for machine learning, and in fact machine learning has been applied such problems with great success.

2.6.1 Classification of Machine Learning

Machine learning implementations are classified into three major categories, depending on the nature of the learning "signal" or "response" available to a learning system which are as follows:

1. **Supervised learning**: When an algorithm learns from example data and associated target responses that can consist of numeric values or string labels, such as classes or tags, in order to later predict the correct response when posed with new examples comes under the category of supervised learning. This approach is indeed similar to human learning under the supervision of a teacher. The teacher provides good examples for the student to memorize, and the student then derives general rules from these specific examples (Figure 2.9).

2. **Unsupervised learning**: Whereas when an algorithm learns from plain examples without any associated response, leaving to the algorithm to determine the data patterns on its own. This type of algorithm tends to restructure the data into something else, such as new features that may represent a class or a new series of uncorrelated values. They are quite useful in providing humans with insights into the meaning of data and new useful inputs to supervised machine learning algorithms.

   As a kind of learning, it resembles the methods humans use to figure out that certain objects or events are from the same class, such as by observing

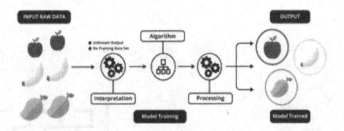

FIGURE 2.9 Supervised learning.

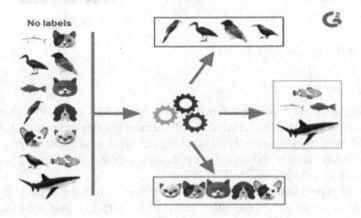

FIGURE 2.10 Unsupervised learning.

the degree of similarity between objects. Some recommendation systems that you find on the web in the form of marketing automation are based on this type of learning (Figure 2.10).

3. **Reinforcement learning**: When you present the algorithm with examples that lack labels, as in unsupervised learning. However, you can accompany an example with positive or negative feedback according to the solution the algorithm proposes comes under the category of reinforcement learning, which is connected to applications for which the algorithm must make decisions (so the product is prescriptive, not just descriptive, as in unsupervised learning), and the decisions bear consequences. In the human world, it is just like learning by trial and error (Figure 2.11).

Errors help you learn because they have a penalty added (cost, loss of time, regret, pain, and so on), teaching you that a certain course of action is less likely to succeed than others. An interesting example of reinforcement learning occurs when computers learn to play video games by themselves.

In this case, an application presents the algorithm with examples of specific situations, such as having the gamer stuck in a maze while avoiding an enemy. The application lets the algorithm know the outcome of actions it takes, and learning occurs while trying to avoid what it discovers to be

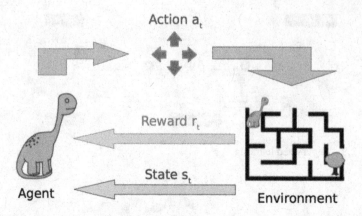

FIGURE 2.11   Reinforcement learning.

dangerous and to pursue survival. You can have a look at how the company Google DeepMind has created a reinforcement learning program that plays old Atari's videogames. When watching the video, notice how the program is initially clumsy and unskilled but steadily improves with training until it becomes a champion.

4. **Semi-supervised learning**: Where an incomplete training signal is given, a training set with some (often many) of the target outputs are missing. There is a special case of this principle known as transduction where the entire set of problem instances is known at learning time, except that part of the targets are missing (Figure 2.12).

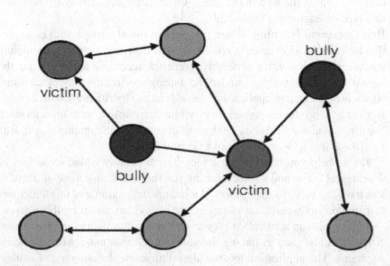

FIGURE 2.12   Semi-supervised learning.

2.6.2 Categorizing on the Basis of Required Output

Another categorization of machine learning tasks arises when one considers the desired output of a machine-learned system:

1. **Classification**: When inputs are divided into two or more classes, and the learner must produce a model that assigns unseen inputs to one or more (multi-label classification) of these classes. This is typically tackled in a supervised way. Spam filtering is an example of classification, where the inputs are email (or other) messages and the classes are spam and not spam.
2. **Regression**: Which is also a supervised problem, a case when the outputs are continuous rather than discrete.
3. **Clustering**: When a set of inputs is to be divided into groups. Unlike in classification, the groups are not known beforehand, making this typically an unsupervised task.

Machine learning comes into the picture when problems cannot be solved by means of typical approaches.

2.7 RECENT TRENDS IN MACHINE LEARNING APPLICATIONS

Recently, there has been a dramatic surge of interest in the era of machine learning, and more people become aware of the scope of new applications enabled by the machine learning approach. It builds a road-map to contact with the device and make the device understandable to response to our instructions and commands.

2.7.1 Traffic Alerts (Maps)

Now, **Google Maps** is probably the most popular app we use whenever we go out and require assistance in directions and traffic. The other day I was traveling to another city and took the expressway, and Maps suggested: "Despite the Heavy Traffic, you are on the fastest route." But, how does it know that? [7] (Figure 2.13).

FIGURE 2.13 Google map logo.

Well, it's a combination of people currently using the service, historic data of that route collected over time, and few tricks acquired from other companies. Everyone using maps is providing their location, average speed, the route in which they are traveling which in turn helps Google collect massive data about the traffic, which makes them predict the upcoming traffic and adjust your route according to it.

2.7.2 SOCIAL MEDIA (FACEBOOK)

One of the most common applications of machine learning is automatic friend tagging suggestions in Facebook or any other social media platform. Facebook uses face detection and image recognition to automatically find the face of the person which matches its database and hence suggests us to tag that person based on DeepFace (Figure 2.14).

Facebook's deep learning project DeepFace is responsible for the recognition of faces and identifying which person is in the picture. It also provides alt tags (alternative tags) to images already uploaded on Facebook. For example, if we inspect the following image on Facebook, the alt-tag has a description (Figure 2.15).

FIGURE 2.14   Face recognition.

FIGURE 2.15   Image recognition.

FIGURE 2.16 Uber icon.

2.7.3 TRANSPORTATION AND COMMUTING (UBER)

If you have used an app to book a cab, you are already using machine learning to an extent. It provides a personalized application which is unique to you. It automatically detects your location and provides options to either go home or office or any other frequent place based on your history and patterns (Figure 2.16).

It uses machine learning algorithm layered on top of historic trip data to make a more accurate **ETA prediction**. With the implementation of machine learning, they saw a 26% accuracy in delivery and pickup.

2.7.4 PRODUCTS RECOMMENDATIONS

Suppose you check an item on Amazon, but you do not buy it then and there. But the next day, you're watching videos on YouTube and suddenly you see an ad for the same item. You switch to Facebook, there also you see the same ad. So how does this happen? (Figure 2.17).

Well, this happens because Google tracks your search history, and recommends ads based on your search history. This is one of the coolest applications of machine learning. In fact, 35% of Amazon's revenue is generated by product recommendations.

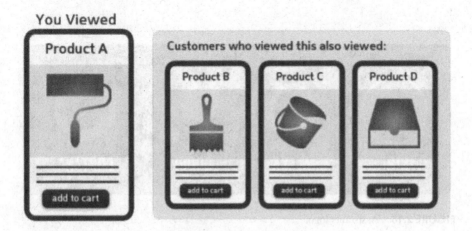

FIGURE 2.17 Products recommendations.

2.7.5 VIRTUAL PERSONAL ASSISTANTS

As the name suggests, virtual personal assistants assist in finding useful information, when asked via text or voice. Few of the major applications of machine learning here are:

- Speech recognition
- Speech to text conversion
- Natural language processing
- Text to speech conversion (Figure 2.18)

All you need to do is ask a simple question like "What is my schedule for tomorrow?" or "Show my upcoming Flights." For answering, your personal assistant searches for information or recalls your related queries to collect info. Recently, personal assistants are being used in chatbots, which are being implemented in various food ordering apps, online training websites, and also in commuting apps.

2.7.6 SELF-DRIVING CARS

Well, here is one of the coolest applications of machine learning. It's here and people are already using it. Machine learning plays a very important role in self-driving cars and I'm sure you guys might have heard about Tesla. The leader in this business and their current artificial intelligence is driven by hardware manufacturer NVIDIA, which is based on an unsupervised learning algorithm (Figure 2.19).

NVIDIA stated that they didn't train their model to detect people or any object as such. The model works on deep learning, and it crowdsources data from all of its vehicles and its drivers. It uses internal and external sensors which are a part of the Internet of Things (IOT). According to the data gathered by McKinsey, the automotive data will hold a tremendous value of $750 billion.

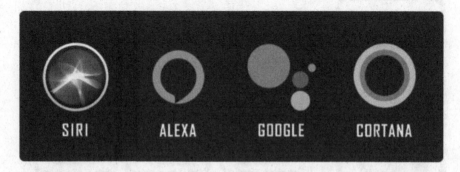

FIGURE 2.18 Assistants logo.

FIGURE 2.19 Tesla logo.

2.7.7 DYNAMIC PRICING

Setting the right price for a good or service is an old problem in economic theory. There are a vast amount of pricing strategies that depend on the objective sought. Be it a movie ticket, a plane ticket or cab fares, everything is dynamically priced. In recent years, artificial intelligence has enabled pricing solutions to track buying trends and determine more competitive product prices (Figure 2.20).

How does Uber determine the price of your ride?

Uber's biggest uses of machine learning comes in the form of surge pricing, a machine learning model nicknamed as "geosurge." If you are late for a meeting and you need to book an Uber in a crowded area, get ready to pay twice the normal fare. Even for flights, if you are traveling in the festive season, chances are prices will be twice the original price.

FIGURE 2.20 Dynamic pricing.

FIGURE 2.21 Google translate logo.

2.7.8 GOOGLE TRANSLATE

Remember the time when you traveled to a new place and you find it difficult to communicate with the locals or finding local spots where everything is written in a different language (Figure 2.21).

Well, those days are gone now. Google's GNMT (Google Neural Machine Translation) is a neural machine learning that works on thousands of languages and dictionaries, uses natural language processing to provide the most accurate translation of any sentence or words. Since the tone of the words also matters, it uses other techniques like POS (point of service) tagging, NER (named entity recognition), and chunking. It is one of the best and most used applications of machine learning.

2.7.9 ONLINE VIDEO STREAMING (NETFLIX)

With over 100 million subscribers, there is no doubt that Netflix is the daddy of the online streaming world. Netflix's speedy rise has all movie industrialists taken aback—forcing them to ask, "How on earth could one single website take on Hollywood?" The answer is machine learning.

The Netflix algorithm constantly gathers massive amounts of data about users' activities like:

- When you pause, rewind, or fast forward
- What day you watch content (TV shows on weekdays and movies on weekends)
- The date and time you watch
- When you pause and leave content (and if you ever come back)
- The ratings given (about 4 million per day), searches (about 3 million per day)
- Browsing and scrolling behavior (Figure 2.22).

FIGURE 2.22 The Netflix logo.

And a lot more. They collect this data for each subscriber they have and use their recommender system and a lot of machine learning applications. That's why they have such a huge customer retention rate.

2.7.10 Fraud Detection

Experts predict online credit card fraud to soar to a whopping $32 billion in 2020. That's more than the profit made by Coca Cola and JP Morgan Chase combined. That's something to worry about. Fraud detection is one of the most necessary applications of machine learning. The number of transactions has increased due to a plethora of payment channels—credit/debit cards, smartphones, numerous wallets, UPI, and much more. At the same time, the amount of criminals have become adept at finding loopholes (Figure 2.23).

FIGURE 2.23 Fraud detection.

Whenever a customer carries out a transaction—the machine learning model thoroughly X-rays their profile searching for suspicious patterns. In machine learning, problems like fraud detection are usually framed as classification problems.

2.7.11 NEWS CLASSIFICATION

It is another benchmark application of a machine learning approach. Why or how? There is a tremendous increase in the volume of information on the web that can be commonly accessed and utilized by anyone. However, every person has his individual interest or choice. So, to pick or gather a piece of appropriate information becomes a challenge to the users from the ocean of this web (Figure 2.24).

Providing that interesting category of news to the target readers will surely increase the acceptability of news sites. Moreover, readers or users can search for specific news effectively and efficiently.

There are several methods of machine learning in this purpose, i.e., support vector machine, naive bayes, k-nearest neighbor, etc. Moreover, there are several news classification software that is available.

2.7.11.1 Video Surveillance

A small video file contains more information compared to text documents and other media files such as audio, images. For this reason, extracting useful information from video, i.e., the automated video surveillance system has become a hot research issue. With this regard, video surveillance is one of the advanced applications of a machine learning approach (Figure 2.25).

The presence of a human in a different frame of a video is a common scenario. In the security-based application, identification of the human from the videos is an important issue. The face pattern is the most widely used parameter to recognize a person [6].

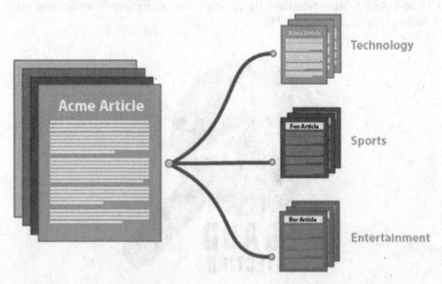

FIGURE 2.24 News classification.

FIGURE 2.25 Video surveillance.

A system with the ability to gather information about the presence of the same person in a different frame of a video is highly demanding. There are several methods of machine learning algorithm to track the movement of human and identifying them.

2.7.11.2 Speech Recognition

Speech recognition is the process of transforming spoken words into text. It is additionally called automatic speech recognition, computer speech recognition, or speech to text. This field benefits from the advancement of machine learning approach and big data (Figure 2.26).

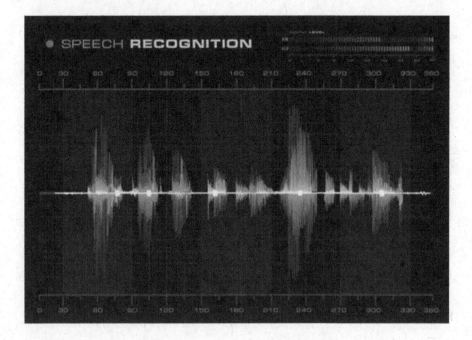

FIGURE 2.26 Speech recognition.

At present, all commercial purpose speech recognition system uses a machine learning approach to recognize the speech. Why? The speech recognition system using machine learning approach outperforms better than the speech recognition system using a traditional method.

2.7.11.3 Robot Control

A machine learning algorithm is used in a variety of robot control system. For instance, recently several types of research have been working to gain control over stable helicopter flight and helicopter aerobatics (Figure 2.27).

In Darpa-sponsored competition, a robot driving for over one hundred miles within the desert was won by a robot that used machine learning to refine its ability to notice distant objects.

2.7.11.4 Author Identification

With the rapid growth of the Internet, the illegal use of online messages for inappropriate or illegal purposes has become a major concern for society. For this regard, author identification is required (Figure 2.28).

Author identification also is known as authorship identification. The author identification system may use a variety of fields such as criminal justice, academia, and anthropology. Additionally, organizations like Thorn use author identification to help end the circulation of child sexual abuse material on the web and bring justice to a child.

FIGURE 2.27 Robot control.

FIGURE 2.28 Author identification.

2.8 WHAT HAPPENS WHEN YOU COMBINE BLOCKCHAIN AND MACHINE LEARNING

We are currently living in the midst of a machine learning revolution with new research and applications coming to fruition. Advancements in ML technology, like deep learning, have changed everything from the recommendations on your Netflix account to how doctors are treating patients before they become ill. It seems like every day there are new developments and new tasks being accomplished by machines previously done by humans. Not too long ago, AI seemed like something only part of a science fiction novel. Today, ML is all around us (Figure 2.29).

FIGURE 2.29 Machine learning.

FIGURE 2.30 Blockchain and machine learning.

Even with all of these current exciting applications of AI, many more will soon be realized in the near future. However, the advent of this new technology also brings unrealistically high expectations. The latest example, machine learning. It's important for people to know where the hype around machine learning ends and where practical applications begin. For me, I see blockchain technology as the enabling infrastructure that will allow machine learning to reach its full potential (Figure 2.30).

2.8.1 INCREASE COMPUTING POWER

In the next few years, our society will undoubtedly be driven by new developments in AI. It will indeed be an exciting world, but it will also require vast amounts of hardware and computational power. For ML to reach these grand visions and deliver on its promises, there needs to be an acceleration of scalable advanced computation available to machine learning tasks.

Currently, huge investments are being made in more and more data centers that are utilizing a traditional CPU-based computing to perform machine learning tasks. A typical CPU unit has between 6–14 cores and can run between 12–28 different threads of command. Usually, these threads will run only on a single data block. So, building more of these CPU data centers will not be enough to meet the growing demand of AI.

However, there is another type of computing that can better satisfy AI's growing demand for computational power, GPU-based computing. A workstation GPU unit can hold between 2,000–3,000 cores and can run 100 or more threads of command with each thread. Usually, these threads will run around 30 blocks of information at the same time. This kind of computing power leads to increased speed and less energy consumption while distributing processing, perfect for machine learning tasks.

Blockchain, or distributed ledger technology (DLT), may provide the computational resources ML needs by utilizing the computing power of machines that hold non-utilized GPU computing power. In some ways, this is what the Bitcoin protocol was designed to do. Part of the Bitcoin protocol requires miners to solve complex mathematical problems that no one computer can solve by itself, as a way to confirm and validate transactions on the blockchain. As the process went on, it evolved and virtual currency was born. If we can tokenize value, can't we also tokenize computing power?

Blockchain-based projects are now working on connecting computers in a peer-to-peer network allowing individual to rent resources out from each other. These resources can be used to complete tasks requiring any amount of computation time and capacity. Today, such resources are supplied by centralized cloud providers which, are constrained by closed networks, proprietary payment systems, and hard-coded provisioning operations (Figure 2.31).

2.8.2 DECREASE COMPUTING COSTS

Every 3.5 months the demand for ML computation is doubling with costs increasing proportionately. Traditional suppliers of computation power, such as Amazon and Microsoft, are using price as a lever to control usage which restricts innovation.

Blockchain-based solutions are now working on building decentralized market-places for GPU computing power that machine learning task need. These projects aim to match a computationally intensive project with connected platform members

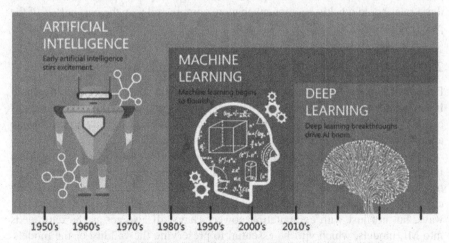

Since an early flush of optimism in the 1950's, smaller subsets of artificial intelligence - first machine learning, then deep learning, a subset of machine learning - have created ever larger disruptions.

FIGURE 2.31   Computing power in different years.

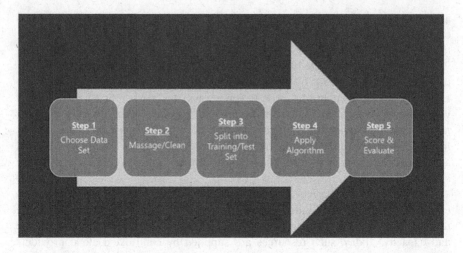

FIGURE 2.32   Steps in machine learning.

who will share their system resources to complete a given task. With DLT, ML innovation can dramatically reduce its cost of computing by accessing the globally distributed GPUs, used by cryptominers, and then make them available to ML companies.

Currently, GPU computing time can be purchased for ~$0.5/hour on multiple cloud platforms compared to ~$0.01–$0.05/hour for CPUs, but despite the higher GPU cost, these types of computation are ~5- to 10-fold cheaper due to vastly shorter runtimes. With blockchain-based projects creating computing power marketplaces, these rates could quickly become much more compressed than the cost curves of the past (Figure 2.32).

2.8.3 IMPROVE DATA INTEGRITY

For any ML model, the presence of accurate and reliable data is central to the intelligent behavior the model produces. This also means accounting for data and application integrity that has unexplainable discrepancies between data incorporated into a model and original records maintained by an engineer.

The very nature of a public blockchain lends itself well to a task such as data integrity. Blockchains create an environment where data is private, immutable, transparent, distributed, and is free to operate without the direction of a sovereign entity. Eventually, public mineable blockchains will be the ML superhighways, but not just with computation power. They will also act as the data feeds into ML models, which will be essential to preserving the validity of the models. Blockchain technologies hold the promise of adding structure and accountability to ML algorithms, as well as the quality and usefulness of the intelligence they produce (Figure 2.33).

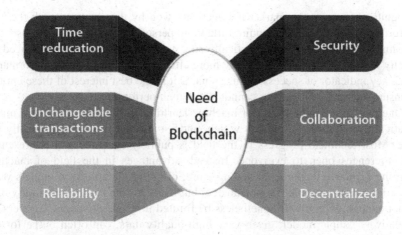

FIGURE 2.33   Need for blockchain.

2.9   MERGING MACHINE LEARNING WITH BLOCKCHAIN

Machine learning relies on vast quantities of data to build models for accurate prediction. A lot of the overhead incurred in getting this data lies in collecting, organizing and auditing the data for accuracy. This is an area that can significantly be improved by using blockchain technology. By using smart contracts, data can be directly and reliably transferred straight from its place of origin. For example, a machine learning model for self-driving trucks would require several hundred terabytes of actual truck driving data. Traditionally, all of the data, like driving speeds, fuel consumption, breaks, etc., would first be collected using different trackers. It would then be sent to a processing facility where auditors would sift through the data to make sure it was authentic before sending it to be processed by data scientists. Smart contracts could, however, improve the whole process significantly by using digital signatures. By using blockchains to ensure the security and ownership of the collected data, we could program smart contracts to directly send the data from the truck driver to the data scientists who would use the data for building machine learning models.

This means that this fusion of blockchain technology and machine learning is a game changer for the self-driving research as it can help create a marketplace for data for research. The finance and insurance industries have a lot to gain as well because together they can be used to design tools to identify and prevent fraud. Using machine learning to improve supply chain solutions can help corporations around the world save billions of dollars every year by reducing wastage and theft.

2.10   BLOCKCHAIN + MACHINE LEARNING: DEMOCRATIZING DATA ACCESS

Having access to superior models over those of your competitors can provide great competitive advantages when using these models as either services or as back-end components to various applications. For example, with something like image recognition services, the market is effectively won by the company with the best performance. These models require little to no person-to-person contact to use, and are simple to hook into programmatically, so it would seem that there is little need for loyalty if a competitor's product were more effective. Therefore, since performance is the key indicator of success in this arena, is it in the best interest of these entities to ensure that their competitors cannot match their performance.

One may assume that superior machine learning ability is a function of mathematical prowess. However, it is widely understood that this is not typically the case. Most technical progress in the field is publicly available and is presented at conferences open to everyone. Instead, advantages in the field of machine learning primarily come from having more, or better, data to train models with. A model can be extremely sophisticated, but if it is trained with low-quality data or not enough data, it will nonetheless be limited in its effectiveness. Conversely, a relatively simple model, given very high-quality data, can often outperform a more complex one that was trained with bad data. Therefore, the ones who will hold the power in the field of machine learning are the ones who have control over, and access to, large amounts of data. Coincidentally, the entities that tend to have the data, such as large tech companies, like Google or Facebook, also tend to have the best researchers and modelers available. They keep private, centralized data repositories that they collect from user data (much of which is voluntarily entered in by users), and can then use these large data-sets to train their cutting-edge models (Figure 2.34).

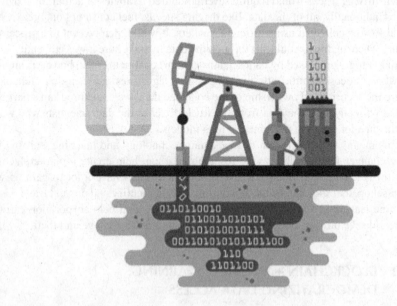

FIGURE 2.34 Accessing information.

Thus enters the potential of blockchain technology in machine learning, primarily in the context of data ownership, collection, and access. If we were to decentralize data collection and allow everyone to access useful data-sets, the competitive moats that these large corporations have would be erased. Given that these data-sets are made from user-contributed activity logs and content, it is only fair that the data is made available to the users who effectively created them.

2.10.1 SYNAPSE AI

This is what the Synapse AI project aims to accomplish with its platform. It is creating a platform in which data contributors are fully aware of the data that they are contributing, and ensures that they are compensated for their contributions (Figure 2.35). For example, users will be able to knowingly contribute their social photos and their tags, or their GPS data, in exchange for compensation. Users of the platform can then pay to access these curated data-sets or trained models in the form of micro-services. The platform aims to create a cyclical economy in which: (1) agents contribute data, (2) data is pooled, (3) models are created using this data, and then (4) agents consume the models. The Synapse ML team hopes that this enables agents in the world to exponentially increase their capabilities by compounding their knowledge of the world through this cyclical process. You can think of this as a sort of automated active learning in which the agent itself autonomously queries for additional information or modeling capabilities. The tokens themselves are used for payments in the platform, for bonding to ensure quality is maintained, and for staking in order to support services (Figure 2.36).

FIGURE 2.35 Synapse AI.

1. Agents
Agents look to fulfill
smart contract and
ad-hoc data queries.

2. Data Pools
Groups of data that
are available for use
to all users.

3. Models
Machine learning
models are created
and trained using
these data pools.

4. Services
These models are
in turn offered back
to agents through
micro services.

FIGURE 2.36 AI tools.

2.10.2 THE BENEFITS OF COMBINING ML AND BLOCKCHAIN

In the next five to ten years, these two technologies are going to be heavily implemented in business. Even now, innovative and tech-savvy industry leaders see significant value in using blockchains with artificial intelligence.

Let's have a look at ways you can use the combination of ML and blockchain to your company's benefit.

- **Enhancing security**: Information in a blockchain is well-protected thanks to inherent encryption. A blockchain is perfect for storing highly sensitive personal data like medical notes or personalized recommendations. Data is what artificial intelligence needs continuously and in high volumes. Currently, experts are busy building algorithms that will allow ML to work with encrypted information without exposing it.

  There's also another angle to security improvements, however. While the blockchain is secure at its base, additional layers and applications are vulnerable (consider breaches of DAO, Bitfinex, etc.). Machine learning will help to improve the deployment of blockchain apps and predict possible system breaches.

- **Untangling the way ML thinks**: Regardless of how great ML is, people won't use it if they don't trust it. One of the issues that has put the brakes on broader adoption of ML is the impossibility to explain decisions made by the computer. With the possibility to record the decision-making process, ML can gain public trust much sooner.

By using the blockchain for artificial intelligence, we can make the way computers think more transparent. A distributed ledger can store every decision made by AI, data point by data point, and make them available for analysis. With a blockchain, you can also be sure that the information is tamper-resistant from recording to examination.

- **Accessing and managing the data market**: This point is tightly connected to enhanced security. Since a distributed ledger can store large amounts of encrypted data and artificial intelligence is able to manage it effectively, new use cases emerge. You can securely store your personal data in the blockchain and sell access to it. As a result, data, model, and ML marketplaces arise.

  Big players like Google, Facebook, and Amazon have access to large volumes of data that can be useful for ML processes, but all of that information is unavailable to others. With a blockchain, smaller companies and startups can challenge the tech giants by accessing the same pool of information and even the same ML potential (we'll talk about the ML marketplace Singularity NET later on).

  Another perk of using artificial intelligence with the blockchain is improving the way we work with data. Computers process encrypted information by going through multiple combinations of characters in search of the correct one to verify a transaction. Similarly to a human hacker, ML learns and sharpens its skills with every successful code crack. But unlike a person, artificial intelligence won't need a lifetime to become an expert. With the right training data, it can happen almost instantly.

- **Optimizing energy consumption**: Data mining is a very energy-consuming process. This is one of the major struggles of the modern world, and Google has proven that machine learning can deal with the issue. Google has managed to reduce energy consumption used for cooling their data centers by 40% by training the DeepMind AI on historical data from thousands of sensors within a data center. The same principle can be used for mining, leading to lower prices for mining hardware.

- **Improving smart contracts**: There are certain technical flaws in the blockchain that can be exploited by hackers (Figure 2.37). This was proven not so long ago. Put simply, smart contracts aren't smart enough. Yet, they're programmed to

FIGURE 2.37   Improving smart contracts.

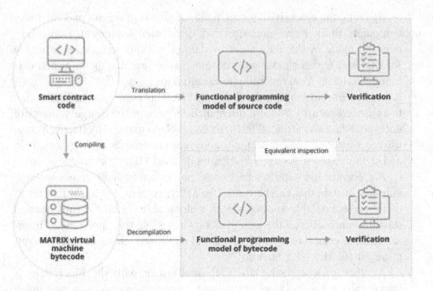

FIGURE 2.38 Formal verification.

release and transfer funds automatically when certain conditions are met. To do that, network consensus must be reached on the blockchain. Smart contract code is public and can be reviewed, so anyone can patiently and thoroughly go through every line of code in search of loopholes. The ML helps to verify smart contracts and predict vulnerabilities that can be exploited.

- **Formal verification of smart contracts**: Plus, artificial intelligence can deal with contracts itself by reviewing conditions and generating and dynamically adjusting smart contracts (Figure 2.38).

2.10.3 Applications of ML and Blockchain

- **Automation in manufacturing**: As a part of the manufacturing procedure, companies are now relying on smart contracts and Bitcoin blockchain-based processes to enable transparency, production, security and compliance checks. Instead of planning traditional fixed machine maintenance schedules, machine learning's predictive algorithms are being used to design flexible plans. Product testing and quality control also have progressively become automated.
- **Food and logistics**: The ML and blockchain are progressively reducing end-to-end supply chain challenges in the food industry by enabling transparency and accuracy. With blockchain coming into play, tracing food sources and management of related financial transactions has become possible.

  Recently, IBM collaborated with Twiga Foods and launched a blockchain-based microfinancing strategy for food vendors. But the mission wouldn't have accomplished without the application of ML techniques. The IBM scientists purchase data from mobile devices, analyzed and then implemented ML algorithms to determine credit scores and predict creditworthiness.

- **Energy and utilities**: In the energy and utilities industry, blockchain is helping in facilitating energy exchanges. For example, IOTA, an energy-based company, has recently implemented blockchain energy production and consumption in a peer-to-peer fashion. Smart energy microgrids are also increasingly becoming a popular way of creation of sustainable energy resources. LO3 Energy, a NY-based company, is also using a blockchain-based innovation for enabling energy generation, conservation, and trading for local communities.

2.10.4 WHAT FIELDS LEVERAGE THE COMBINATION OF ARTIFICIAL INTELLIGENCE AND THE BLOCKCHAIN?

All of the above are theoretical benefits of combining a blockchain and AI. Now, let's move to real-life cases. Different industries are already testing the waters of using blockchain and ML together in one project. It's still too early to pop the champagne, but the first tries look really promising.

Porsche is officially the first automaker to test the blockchain in vehicles. It partnered with the German startup XAIN to implement blockchain technology in its brand-new sports car. Using their smartphones, drivers will be able to record traffic data, which has been received from connected vehicles, on a blockchain. The solution will also allow owners to grant temporary access to a car and receive notifications about who accesses it, where, and when. And let's not forget about the increased security and auditable and usable data for predictive maintenance and autonomous driving.

2.11 ADDRESSING BLOCKCHAIN AND ML LIMITATIONS

Although the blockchain can do some amazing things, there are still limitations to this relatively new technology. A *Medium* article on these limitations listed some of the blockchain's limitations. These included significant energy consumption during the mining process, scalability problems, and some security vulnerabilities. Other problems include privacy, efficiency, hardware, and a lack of talent that understands the real power of this technology. Enter AI. Integrating machine learning to the blockchain can efficiently power the blockchain in a cost-effective way. Plus, it adds virtual talent to improve on what this technology can do.

Also, blockchain can help ML progress. As the article stated, the blockchain can help ML explain itself. "The ML black-box suffers from an explainability problem. Having a clear audit trail can not only improve the trustworthiness of the data as well as of the models but also provide a clear route to trace back the machine decision process." Additionally, the blockchain has already proven it can add efficiency and speed to transactions. Therefore, it can do the same for AI, propelling it to learn at a faster rate. The blockchain's model offers a good benchmark for organizing information in a more efficient way.

2.12 CONCLUSION

The combination of blockchain technology and machine learning is still a largely undiscovered area. Even though the convergence of the two technologies has received its fair share of scholarly attention, projects devoted to this groundbreaking combination are still scarce.

When people with access to the highest quality information concerning ML, like Elon Musk, are saying that ML could be the biggest existential threat in existence, giving this power to one single entity isn't the best idea. Decentralizing ML and letting it be designed and controlled by a large network through open-source programming is probably the safest approach to create super intelligence.

Putting the two technologies together has the potential to use data in ways never before thought possible. Data is the key ingredient for the development and enhancement of ML algorithms, and blockchain secures this data, allows us to audit all intermediary steps ML takes to draw conclusions from the data and allows individuals to monetize their produced data.

The ML can be incredibly revolutionary, but it must be designed with utmost precautions—blockchain can greatly assist in this. How the interplay between the two technologies will progress is anyone's guess. However, its potential for true disruption is clearly there and rapidly developing.

REFERENCES

1. Dhingra, P., Gayathri, N., Kumar, S. R., Singanamalla, V., Ramesh, C., & Balamurugan, B. (2020). Internet of Things–based pharmaceutics data analysis. In *Emergence of Pharmaceutical Industry Growth with Industrial IoT Approach* (pp. 85–131). Academic Press, Cambridge, MA.
2. Nagasubramanian, G., Sakthivel, R. K., Patan, R., Gandomi, A. H., Sankayya, M., & Balusamy, B. (2018). Securing e-health records using keyless signature infrastructure blockchain technology in the cloud. *Neural Computing and Applications*, *32*(3), 639–647.
3. Muthuramalingam, S., Bharathi, A., Gayathri, N., Sathiyaraj, R., & Balamurugan, B. (2019). IoT Based Intelligent Transportation System (IoT-ITS) for global perspective: A case study. In *Internet of Things and Big Data Analytics for Smart Generation* (pp. 279–300). Springer, Cham.
4. Rana, T., Shankar, A., Sultan, M. K., Patan, R. & Balusamy, B. (2019, January). An intelligent approach for UAV and Drone Privacy Security using Blockchain Methodology. In *2019 9th International Conference on Cloud Computing, Data Science & Engineering (Confluence)* (pp. 162–167). IEEE.
5. Rizwan, P., Babu, M. R., Balamurugan, B., & Suresh, K. (2018, March). Real-time big data computing for internet of things and cyber physical system aided medical devices for better healthcare. In *2018 Majan International Conference (MIC)* (pp. 1–8). IEEE.
6. Khari, M., Garg, A. K., Gandomi, A. H., Gupta, R., Patan, R., & Balusamy, B. (2019). Securing data in Internet of Things (IoT) using cryptography and steganography techniques. *IEEE Transactions on Systems, Man, and Cybernetics: Systems, 50*(1), 73–80.
7. Karthikeyan, S., Patan, R., & Balamurugan, B. (2019). Enhancement of security in the Internet of Things (IoT) by using X. 509 authentication mechanism. In *Recent Trends in Communication, Computing, and Electronics* (pp. 217–225). Springer, Singapore.

WEB RESOURCES

1. www.dataversity.net/blockchain-trends-in-2019/
2. www.computerworld.com/article/3427960/blockchain-and-cryptocurrency-
3. https://blockonomi.com/four-trends-blockchain-emerging-tech-2019/
4. https://thefintechtimes.com/blockchain-trends-2019/
5. www.tandfonline.com. › doi › full
6. www.mckinsey.com › industries › high-tech › our-insights › howto
7. www.grandviewresearch.com › industry-analysis › blockchain-tech
8. www.getsmarter.com › blog › market-trends ›
9. www.graymentechnologies.com › blog › top-7-blockchain-tech
10. www.deccanchronicle.com › technology › in-other-news › 5-trends
11. www.edureka.co/blog/machine-learning-applications/
12. https://medium.com/app-affairs/9-applications-of-machine-learning
13. www.ubuntupit.com/top-20-best-machine-learning-applications-in-real-world/
14. https://medium.com/@Intersog/what-happens-when-you-combine-blockchain-and-machine-learning
15. https://hackernoon.com/3-ways-blockchain-will-unleash-the-full-potential-of-machine-learning
16. www.blockchain-council.org/blockchain/can-blockchain-and-machine-learning-work-together/
17. https://tokensale.synapse.ai/r/59860
18. https://espeoblockchain.com/blog/decentralized-ai-benefits/
19. www.finextra.com › blogposting › un-block-your-business-10-com
20. www.techcircle.in › 2018/05/11 › ai-blockchain-combine-to-spur.
21. https://robots.net › fintech › cryptocurrency › cryptocurrency-jobs
22. www.alternativky.com › iot-trends-challenges-and-future-scope
23. https://imarticus.org/what-is-difference-between-blockchain-and-machine-learning/
24. https://books.google.co.in › books
25. https://developer.ibm.com › articles › introduction-watson-studio
26. www.datasciencecentral.com › profiles › blogs › blockchain-and-art...
27. www.upgrad.com › Home › Data › Machine Learning
28. www.analyticsinsight.net › big-data-deep-learning-and-blockchain
29. www.computer.org › csdl › magazine › 2018/09 › mco2018090048
30. https://dataconomy.com › 2017/05 › blockchains-data-scientist-dream

3 Blockchain Databases 1

S. Dhivya, S. V. Evangelin Sonia, and P. Suvithavani

CONTENTS

3.1   DATABASE

A traditional database is a data structure used for storing information. A database stores information in data structures called tables. A regular database is amassed in information components called a table. Tables contain fields, which define the type of record that store data called attributes. Each field contains columns that describe the field and rows which define a record stored in a database (Figure 3.1).

A database can be modified, managed and controlled by a single user called an administrator. The database always has a user that functions as a DB admin and that user has complete control of the database. This user can create, delete, modify and change any record stored in a database. They can also perform administration on the database like optimizing performance and managing its size to more manageable levels. A large database tends to slow performance, so admins can run optimization methods to improve performance.

A database stores information in data structures called tables. This includes data that can be queried to gather insights for structured reporting used by entities to support business, financial, and management decisions. The government also makes use of databases to store large sets of data which scale to millions of records (Figure 3.2).

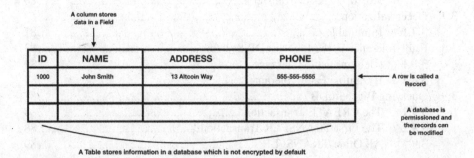

FIGURE 3.1   Shows a sample database.

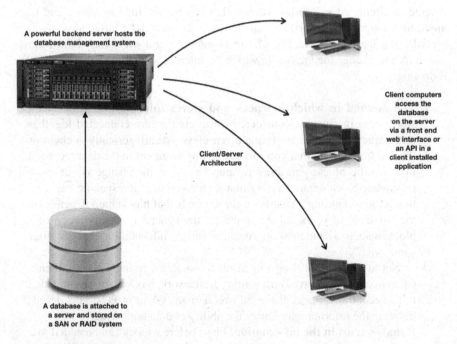

FIGURE 3.2   Database topology.

3.2   BLOCKCHAIN

A blockchain is really a database since it is an advanced record that stores data in information structures called blocks [1]. A blockchain stores information in uniform-sized blocks; each structure contains the hashed data from the earlier block to give cryptographic security. The hashing uses SHA256, which is a one-way hash function. This hashed information is the data and digital signature from the previous block, and the hashes of previous blocks that goes all the way back to the very first block delivered in the blockchain called a "beginning block." That data is gone through a hash function that at that point focuses to the location of the earlier block. A blockchain data structure is an example of a Merkle tree, which is used as an efficient way to verify data [2] (Figure 3.3).

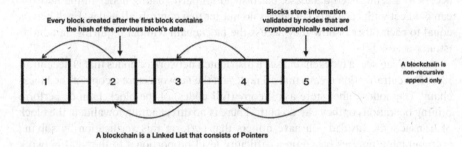

FIGURE 3.3   Structure of blockchain.

A suite of circulated record innovations that can be modified to record and track anything of significant worth. It can be used in financial transactions, medical records, and land titles, etc. [3]. The motivation behind why blockchain innovation stands to change the manner in which we interact with one another include the following:

1. **The method in which it tracks and stores information**: Blockchain stores data in clusters, considered obstructs that are connected together in a sequential manner to frame a ceaseless line allegorically, a chain of blocks. In the event that you roll out an improvement to the data recorded in a specific block, you don't revamp it; rather, the change is put away in another block demonstrating that x changed to y at a specific date and time. Sound familiar? That is on the grounds that blockchain depends on the hundreds of years old technique for the general monetary record. So blockchain is a "biocidal approach to follow information changes after some time."

   Not at all like the deep-rooted record strategy, initially a book then a database document saved on a solitary framework, blockchain was intended to be decentralized and dispersed over a huge system of PCs. This decentralizing the information reduces the ability of data tampering.

2. **It makes trust in the information**: First, before a block can be added to a chain, a couple of things need to occur. Initial a cryptographic riddle must be solved, along these lines making the block. Second, the computer that solves the puzzle shares the solution to all the other computers on the network. This is called proof-of-work. Third, the network verifies the proof of work. If it is fact, the block will be added to the chain.

   The combination of these complex math puzzles and verification by many computers ensures that we can trust each and every block of the chain because the network does the trust building for us. We now have the opportunity to interact directly with our data in real-time. Furthermore, that carries us to the third reason blockchain innovation is such a distinct advantage. That is the blockchain has no more intermediaries (Figure 3.4).

A blockchain uses a peer-to-peer (P2P) network architecture. It does not require access to a centralized database, but instead all participating nodes in the network can connect with each other. There is no master that controls all nodes. Each peer is equal to each other in how they access the blockchain without requiring an administrator access [4].

As you can see, a blockchain uses a distributed network of nodes that is decentralized. Decentralization means that all nodes on the network store a copy of the blockchain. The nodes either store a full copy (full nodes) of the blockchain or perform mining operations or they can do both. There is no direct admin to validate the block of transactions. Instead you have miners that perform this verification by solving cryptographic puzzles based on a difficulty level proportional to the total network hashing power available.

P2P Architecture

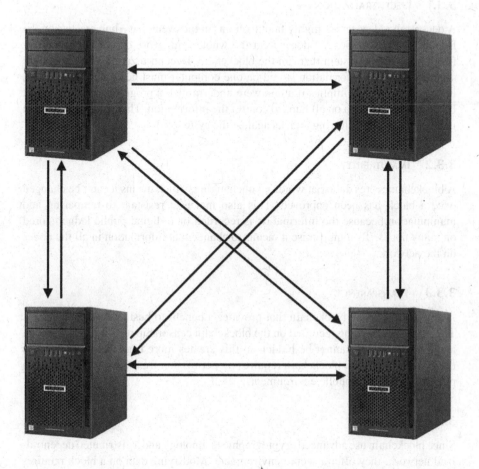

**All nodes can communicate directly with each other
and store a copy of the blockchain**

FIGURE 3.4 Blockchain topology.

When the block has been added to the blockchain, the data is changeless and straightforward to all. Blockchain transactions are non-recursive, meaning they cannot be repeated once validated in a block. A blockchain is highly fault tolerant since if one or more nodes are down, there will always be other nodes available that will run the blockchain. Another advantage of decentralization is that it can be permissionless and trustless, allowing people who don't know or [5] each other to transact. What the blockchain does is provide that trust through transparency by recording the transaction and providing a cryptographically secure way to exchange value.

3.3   ADVANTAGES TO WHY WE USE A BLOCKCHAIN

3.3.1   DECENTRALIZATION

A decentralized system is highly fault tolerant. In the event that a hub crashes on the Bitcoin systems suppose, it doesn't cut the whole framework down. There are different hubs on the system that run the blockchain. Decentralization also adds more security since the information stored on one computer must be copied to all nodes in the network. This implies if a hub were undermined, a programmer should most likely change the data on all hubs to control the information. This has proven to be a good safeguard in deterring attacks against the system [6].

3.3.2   IMMUTABILITY

A blockchain stores data that winds up unchanging, which means it can't be changed once a block has been approved. This also makes it resistant to tampering and manipulation because the information is recorded on a digital public ledger stored on many nodes. To compromise it means to change that information in all the nodes on the network.

3.3.3   TRANSPARENCY

A key feature of the blockchain that provides a benefit to business is transparency. This makes everything recorded on the blockchain censorship resistant. Information about a transaction cannot be hidden so this creates more trust and adds value to the system. Using the blockchain requires no permission from anyone, it is an open platform for all in a public environment.

3.3.4   SECURITY

Since blockchain use advanced cryptographic technology and a distributed decentralized network, they offer a secure environment. Modifying data on a block requires expending plenty of compute resources. It also is not ideal because it requires changing the data on all nodes on the network. This is what deters attacks since it is more costly than mining blocks for rewards. This is a feature to help protect the blockchain from rogue miners and hackers.

3.4   THE PROBLEMS WITH BLOCKCHAIN

3.4.1   ENERGY CONSUMPTION

First and foremost, the compute resources like Bitcoin to run a blockchain expends large amounts of electricity. This is part of the protocol required to process transactions in the proof-of-work algorithm. All the energy is used by the miners in order to solve cryptographic puzzles to validate blocks. The amount of energy consumed increases with the level of difficulty increase that is related to more hashing power from compute resources. The more nodes you have mining, the greater the computational effort required to validate a block of transactions. This requires plenty of energy consumed. The whole Bitcoin network has been estimated to consume the same amount of electricity as small country like Haiti or Denmark [7]. The proof-of-work (PoW) is used to solve the problem. This algorithm is used to confirm transactions and produce new blocks to the chain. With PoW, miners compete against each other to complete transactions on the network and get rewarded.

3.4.2 SCALABILITY

Blockchains do not scale well when it comes to high volume transactions. Because of the fixed block size, there are issues with expanding transaction volume. The delays also affect transaction velocity, where most blockchains cannot process more than 15 transactions per second. Scaling solutions have become the focus of many projects to optimize performance to handle more transactions and increase processing time. If claims to 1 million transactions per second are proven on the blockchain (not yet as of this writing), then that can significantly disrupt the rest of the industry.

3.4.3 SIZE

An issue with most databases, including blockchains, is their size. When they get bigger, they consume more space for storage and this makes them slow down. Bitcoin'sblockchain is already >100 GB, while Ethereum'sblockchain size has surpassed 1 TB (as of this writing). It's not just a storage issue for nodes, but a network as well. With larger blockchain sizes, it takes much longer to copy them to new nodes on the network. It can take several hours to days depending on the network bandwidth. The bigger blockchain size requires more data transfer capacity to transmit to another hub. This affects new nodes or nodes that go back online and have not been updated in a long time.

3.4.4 HIGH TRANSACTION FEES

The expenses to process transactions are another issue which Bitcoin confronted. At whatever point the interest is high, exchange charges additionally go up to profit excavators. Keeping transaction charges low or evacuating transaction expenses is a test for blockchain architects. With high transaction expenses, clients are stopped from utilizing the system. When scaling issues tackle the issues with transaction speed and volume, progressively sensible charges ought to be connected.

3.4.5 INTEROPERABILITY

This is at present an issue since, in contrast to conventional databases, each blockchain is especially its own environment. There are protocols that aim to make blockchains interoperate with each other. For instance, to enable clients to move an incentive from Bitcoin to another blockchain like Ethereum requires the utilization of an advanced trade. Developers are finding ways to make dissimilar blockchains interoperable to make the transfer of value much simpler.

Databases are best for enterprise networks because of their stability. For instance, to enable clients to move an incentive from Bitcoin to another blockchain like Ethereum requires the utilization of an advanced trade. The best 500 organizations on Forbes utilize databases that run top of the line frameworks that manage huge volumes of information. Databases can scale to millions of records and process thousands of transactions per second very easily. For frameworks that manage high volume traffic, similar to retail, a database is as yet the best arrangement. The stock market is better off with a database that can quickly store information and allow instant retrieval without the need for miners to validate the data. A blockchain does not need to store large amounts of numerical data used in analytical processing. A database can store this information much better and procedure it quicker also since it doesn't require numerous hubs to run each bit of information. You also don't need to encrypt or hash every piece of data you store in a database. By default, databases are unencrypted because encryption adds a lot of overhead in a live database. Being permissioned is the security feature in a traditional database [3]. A database that is archived can be encrypted, however.

Databases have proven their reliability for storing information and providing quick queries to retrieve data for reports and analytical purposes. Unstructured data is another thing that does not require a blockchain, these are more suitable for database management systems. Data that does not need to trust verification to be used, like the number of items sold by a store at the end of the day is best recorded on a database. It is also more costly to use a blockchain for something as simple as private bookkeeping information, since a standalone database is more efficient. Personal information that only a certain company needs to know like social security and medical records are best stored in databases. This data can be utilized with open confirmation frameworks that can depend on a blockchain. The personal information can be obscured but verified on the blockchain based on public key cryptography.

A database is ideal for:

- Data that need continuous updating, like monitoring and sensors
- Fast online transaction processing
- Confidential information (non-transparent to the public)
- Financial data from markets that require fast processing
- Data that does not require verification
- Standalone applications that store data
- Relational data

The requirements for blockchains are to establish trust and transparency. It is simply a digital public ledger which allows everyone access to information. In this case it can help with validating information from B2B business-to-business transactions related to supply chain, distribution, and inventory. Transparency can help with industries like advertising to minimize fraud by building more verification of an advertiser's company and the source of ad spends. Blockchains while not for large scale data records can be implemented more for validating information. Bitcoin is the first successful implementation of a blockchain, and it works well as a system for transferring value and validating payments in transactions. Bitcoin's success is that it also addresses the double spend problem in digital payment systems that would have allowed users to spend the same coin more than once. Bitcoin implements a protocol that validates transactions using confirmations based on a chronological order with timestamps and the user's funds that are available. This helps to prevent double spending by not allowing the system to process transactions simultaneously; they will always be done in chronological order.

Some projects are exploring blockchains for permissioned systems like those used in voting stations. It makes a lot of sense on paper since a blockchain can verify both the identity and the vote made by a person. The purpose is to prevent cheating, so blockchains really aim to enforce fairness in trustless and permissionless systems and likewise in trusted and permissioned systems as well. In the case of the latter, some blockchains don't require cryptocurrency or mining, like in enterprise blockchains. These are a new class of systems that use blockchain technology in a private and permissioned environment, and sometimes integrated with databases to form a hybrid system.

One thing long time database administrators will notice is that blockchains are non-relational. You cannot create joins on different blockchains and relate data. This is a major difference between the two, so when information needs to be relational a blockchain will not be suitable for it.

Other blockchains implement what are called "Smart Contracts" like on the Ethereum network. These are much like using stored procedures in a database, in which triggers can be used to execute code to process a transaction. In Ethereum's network, a smart contract execute as bytecode on all nodes in the network. Ethereum and other cryptocurrency like EOS and NEO use blockchains as a platform for their smart contract ecosystem. This is another case of how blockchain use can contrast from traditional databases.

A blockchain is ideal for:

- Monetary exchanges
- Transfer of worth
- Verification of trusted data (identity, reputation, credibility, integrity, etc.)
- Public key verification
- Decentralized applications (DApps)
- Voting frameworks

3.5 BIGCHAINDB

BigchainDB is a versatile blockchain database. It's intended to blend the best of two universes: the traditional distributed database world and the conventional blockchain world [8].

BigchainDB begins with a conventional distributed database (at first RethinkDB), which has the qualities of:

- **scale** (throughput, capacity, low latency)
- **queryability**

From that, we engineered in blockchain characteristics:

- **decen.tralized** (no single entity owns or controls it)
- **immutable** (tamper-resistance)
- **assets** (owns the asset if they owns the private key, i.e., blockchain-style permissioning)

3.6 KEY CONCEPTS OF BIGCHAINDB

- Transaction model
- Visualization of transaction model
- Asset
- Input
- Output
- Metadata
- Transaction ID

3.6.1 TRANSACTION MODEL

The main thing to comprehend about BigchainDB is the way we structure our information. Traditional SQL databases structure data in tables. NoSQL databases use other formats to structure data such as JSON and key-values, as well as tables. At BigchainDB, we structure data as assets. We believe anything can be represented as an asset. An asset can characterize any physical or digital object that you can think of like a car, a data set or an intellectual property right [9].

These advantages can be enlisted on BigchainDB in two different ways. (1) By users in CREATE transactions. (2) Updated (or refreshed) to different clients in TRANSFER exchanges. Traditionally, people design applications focusing on business processes (e.g., apps for booking and processing client orders, apps for tracking delivery of products, etc.). At BigchainDB, we don't concentrate on procedures rather on resources (e.g., a customer request can be a benefit that is then followed over its whole lifecycle). This switch in perspective from a process-centric towards an asset-centric view influences much of how we build applications.

3.6.2 VISUALIZATION OF TRANSACTION MODEL

This infographic will help to understand what CREATE and TRANSFER exchanges are and what the individual segments of a transaction insists (inputs, yields, resources, metadata and so forth). Let's see a basic genuine model, for example, Martina carefully enrolls her bike on BigchainDB in a CREATE exchange. After some time, she moves this bike to Stefan in a transaction (Figure 3.5) [9].

3.6.3 ASSET

An asset constitutes any physical or computerized object. It can be a physical object like a car or a house. Or it can be a digital object like a customer order or an air mile. An asset can have one or multiple owners, but it can also be its own owner. Think of an autonomous car or an Internet of Things (IoT) sensor that does transactions

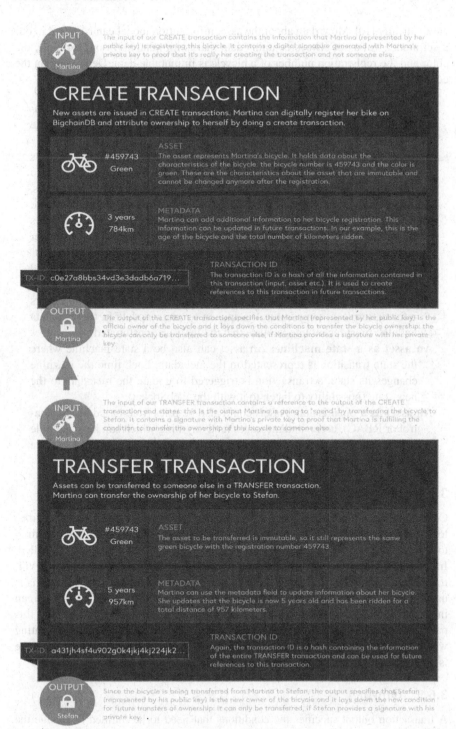

FIGURE 3.5   Shows the virtualization of the transaction model.

automatically [10]. More data about the asset information model can be found in our asset model. An asset always contains data that is immutable. In our example, the color and the registration number of a bicycle is immutable data. Depending on the context, an asset can represent many different things.

An asset as a claim: An asset can represent an ownership claim for a particular object, for example, it represents a claim that User ABC owns the bicycle with the number XYZ. This can be valid for any type of ownership.

An asset as a token: An asset can also represent a token. BigchainDB supports divisible assets. This means, multiple assets can be issued and attributed to one overarching asset. This can for example be intriguing for token dispatches.

An asset as a versioned document: An asset can likewise be a formed record with the adaptation expressed in the metadata field. The adaptation of this report can be refreshed on a nonstop premise. Each time there is another adaptation of the archive, it could be reflected in the metadata.

An asset as a period arrangement: An asset can likewise speak to a period arrangement of information. For example, an IoT sensor records its own information. The IoT sensor is the asset and each accommodation of its information (e.g., temperature) is spoken to as an update in the metadata with the most recent temperature that the IoT sensor estimated.

An asset as a state machine: An asset can also be a state machine where the state transition is represented in the metadata. Each time the machine changes its state, a transaction is triggered to update the metadata to the new state (possibility to listen to it with the WebSocket).

An asset as a permission: Assets could also be roles (role-based access control or RBAC), users, and messages (and anything which can have multiple instances in a scenario—vehicles, reports, and so on). As you can see, there are almost no limits with respect to what an asset can represent.

3.6.4 INPUT

Conceptually, an input is a pointer to an output of a previous transaction. It specifies to whom an asset belonged before and it provides a proof that the conditions required to transfer the ownership of that asset (e.g., a person needs to sign) are fulfilled. In a CREATE transaction, there is no previous owner, so an input in a CREATE transaction simply specifies who the person is that is registering the object (this is usually the same as the initial owner of the asset). In a TRANSFER transaction, an input contains a proof that the user is authorized to spend (transfer or update) this particular output. In practical terms, this means that with the input, a user is stating which asset (e.g., the bike) should be transferred. He also demonstrates that he or she is authorized to do the transfer of that asset.

3.6.5 OUTPUT

A transaction output specifies the conditions that need to be fulfilled to change the ownership of a specific asset. For instance: to transfer a bicycle, a person needs to sign the transaction with his or her private key. This also implicitly contains the information that the public key associated with that private key is the current owner of the asset.

The transaction can also have multiple outputs. These are called divisible assets. The output can also contain complex conditions (e.g., multiple signatures of multiple people) to acquire ownership.

3.6.6 METADATA

The metadata field allows users to add additional data to a transaction. This can be any type of data, like the age of a bicycle or the kilometers driven [11]. The good thing about the metadata is that it can be updated with every transaction. In contrast to the data in the asset field, the metadata field allows to add new information to every transaction.

3.6.7 TRANSACTION ID

The ID of a transaction is a unique hash that identifies a transaction. It contains all the information about the transaction in a hashed way.

3.6.8 GETTING STARTED

Begin by creating an object of class BigchainDB:

```
In [1]: frombigchaindb_driverimportBigchainDB
In [2]: bdb_root_url='https://example.com:9984'#Use YOUR
BigchainDB Root URL here
```

If the BigchainDB node or cluster doesn't require authentication tokens, you can do:

```
In [3]: bdb=BigchainDB(bdb_root_url)
```

If it *does* require authentication tokens, you can do put them in a dict like so:

```
In [4]: tokens={'app_id':'your_app_id','app_key':'your_app_key'}
In [5]: bdb=BigchainDB(bdb_root_url, headers=tokens)
```

3.6.9 DIGITAL ASSET DEFINITION

As an example, let's consider the creation and transfer of a digital asset that represents a bicycle:

```
In [6]: bicycle={
   ...: 'data':{
   ...: 'bicycle':{
   ...: 'serial_number':'abcd1234',
   ...: 'manufacturer':'bkfab',
   ...:},
   ...:},
   ...:}
   ...:
```

Suppose a bike belongs to Alice and that it will be transferred to Bob. In general, use any dictionary for the data property.

3.6.10   METADATA DEFINITION (*OPTIONAL*)

You can *optionally* add metadata to a transaction. Any dictionary is accepted.
   For example:

```
In [7]: metadata={'planet':'earth'}
```

Asset Creation
   Create the digital asset. First, prepare the transaction:

```
In [10]: prepared_creation_tx=bdb.transactions.prepare(
....: operation='CREATE',
....: signers=alice.public_key,
....: asset=bicycle,
....: metadata=metadata,
....:)
....:
```

The prepared_creation_tx dictionary should be similar to

```
In [11]: prepared_creation_tx
Out[11]:
{'asset': {'data': {'bicycle': {'manufacturer': 'bkfab',
   'serial_number': 'abcd1234'}}},
'id': None,
'inputs': [{'fulfillment': {'public_key':
'7rr59L64LvvwMKr8dtieotdugk13oHkosFFoTuXCjzz6',
   'type': 'ed25519-sha-256'},
   'fulfills': None,
   'owners_before': ['7rr59L64LvvwMKr8dtieotdugk13oHkosFFoTuXCj
      zz6']}],
'metadata': {'planet': 'earth'},
'operation': 'CREATE',
'outputs': [{'amount': '1',
 'condition': {'details': {'public_key': '7rr59L64LvvwMKr8
   dtieotdugk13oHkosFFoTuXCjzz6',
   'type': 'ed25519-sha-256'},
 'uri': 'ni:///sha-56;e2C00fzLhNG_nLwWVmZOyGbIQ61qDCduYZy7HKPr2
   3c?fpt=ed25519-sha-256&cost=131072'},
 'public_keys': ['7rr59L64LvvwMKr8dtieotdugk13oHkosFFoTuXCj
   zz6']}],
'version': '2.0'}
```

The transaction now needs to be fulfilled by signing it with Alice's private key:

```
In [12]: fulfilled_creation_tx=bdb.transactions.fulfill(
....: prepared_creation_tx, private_keys=alice.private_key)
....:

In [13]: fulfilled_creation_tx
```

```
Out[13]:
{'asset': {'data': {'bicycle': {'manufacturer': 'bkfab',
   'serial_number': 'abcd1234'}}},
'id': 'bfe679e84f528410270fe67d309f19d03afe09970650140621d
   aedfe6f546598',
'inputs': [{'fulfillment': 'pGSAIGXrEQNtYgyzOBswGHnoWgbU
   WFWVE89Kez 25zti91PTzgUDY8sy8E9FtHwO9-nab89wBspemhBCshbbRR
   JqLlqilTQ7Hb JZYMfr8_mawfTDre31pJEgLSrqFU-clmJjkrwUO',
   'fulfills': None,
   'owners_before': ['7rr59L64LvvwMKr8dtieotdugk13oHkosFFoTuXCj
   zz6']}],
'metadata': {'planet': 'earth'},
'operation': 'CREATE',
'outputs': [{'amount': '1',
   'condition': {'details': {'public_key':
   '7rr59L64LvvwMKr8dtieotdugk13oHkosFFoTuXCjzz6',
   'type': 'ed25519-sha-256'},
   'uri': 'ni:///sha-256;e2C00fzLhNG_nLwWVmZOyGbIQ61qDCduYZy7HK
     Pr23c?fpt=ed25519-sha-256&cost=131072'},
   'public_keys': ['7rr59L64LvvwMKr8dtieotdugk13oHkosFFo
   TuXCjzz6']}],
'version': '2.0'}
```

And sent over to a BigchainDB node:

```
>>>sent_creation_tx=bdb.transactions.send_commit
   (fulfilled_creation_tx)
```

Note that the response from the node should be the same as that which was sent:

```
>>>sent_creation_tx==fulfilled_creation_tx
True
```

Notice the transaction **id:**

```
In [14]: txid=fulfilled_creation_tx['id']

In [15]: txid
Out[15]: 'bfe679e84f528410270fe67d309f19d03afe09970650140621
   daedfe6f546598'
```

Check if the Transaction was sent successfully

After a couple of seconds, we can check if the transaction was included in a block.

```
# Retrieve the block height
>>>block_height=bdb.blocks.get(txid=signed_tx['id'])
```

This will return the block height containing the transaction. If the transaction is not in any block then None is returned. If it is None, it can have different reasons for example the transaction was not valid or is still in the queue and you can try again later. If the transaction was invalid or could not be sent an exception is raised.

If we want to see the whole block we can use the block height to retrieve the block itself.

```
# Retrieve the block that contains the transaction
>>>block=bdb.blocks.retrieve(str(block_height))
```

3.6.10.1　Asset Transfer

Imagine some time goes by, during which Alice is happy with her bicycle, and one day she meets Bob, who is interested in acquiring her bicycle. The timing is good for Alice as she wanted to get a new bicycle [12].

To transfer the bicycle (asset) to Bob, Alice must consume the transaction in which the Bicycle asset was created.

Alice could retrieve the transaction:

```
>>>creation_tx=bdb.transactions.retrieve(txid)
```

or simply use fulfilled_creation_tx :

```
In [16]:creation_tx=fulfilled_creation_tx
```

In order to prepare the transfer transaction, we first need to know the id of the asset we'll be transferring. Here, because Alice is consuming a CREATE transaction, we have a special case in that the asset id is NOT found on the asset itself, but is simply the CREATE transaction's id:

```
In [17]:asset_id=creation_tx['id']
In [18]:transfer_asset={
....:  'id':asset_id,
....:}
....:
```

Let's now prepare the transfer transaction:

```
In [19]: output_index=0
In [20]: output=creation_tx['outputs'][output_index]
In [21]: transfer_input={
....: 'fulfillment':output['condition']['details'],
....: 'fulfills':{
....: 'output_index':output_index,
....: 'transaction_id':creation_tx['id'],
....:},
....: 'owners_before':output['public_keys'],
....:}
....:
In [22]: prepared_transfer_tx=bdb.transactions.prepare(
....: operation='TRANSFER',
....: asset=transfer_asset,
....: inputs=transfer_input,
....: recipients=bob.public_key,
....:)
....:
fulfill it:
In [23]: fulfilled_transfer_tx=bdb.transactions.fulfill(
....: prepared_transfer_tx,
....: private_keys=alice.private_key,
....:)
....:
```

The fulfilled_transfer_tx dictionary should look something like:

```
In [24]: fulfilled_transfer_tx
Out[24]:
{'asset': {'id': 'bfe679e84f528410270fe67d309f19d03afe09
    970650140621daedfe6f546598'},
'id': 'fb745cf6b645f5a20a82a41adef835ce66913cc61dd01b2830085d
    7776855520',
'inputs': [{'fulfillment': 'pGSAIGXrEQNtYgyzOBswGHno
    WgbUWFWVE89Kez25zti91PTzgUBuHqMbJO6C2vm9aBcaz90CBXpg
    PZ-FQPrFLeGURN77an7EjYXGvLGeUhf-fTB9uZvpYs6utzhZPt
    24rdSJJXQC',
    'fulfills': {'output_index': 0,
    'transaction_id':
    'bfe679e84f528410270fe67d309f19d03afe09970
    650140621 daedfe6f546598'},
    'owners_before': ['7rr59L64LvvwMKr8dtieotdugk13oHkos
    FFoTuXCjzz6']}],
'metadata': None,
'operation': 'TRANSFER',
'outputs': [{'amount': '1',
    'condition': {'details': {'public_key': '5RaeQAZEEQMFCLBEDr
    Ktk5p Lue4SuVkdw9Xb55ucpvQ6',
    'type': 'ed25519-sha-256'},
    'uri': 'ni:///sha-256;UKI3FbYGLQhOeu3QkZNuoWrjxGugzG4q7LxnhV
    qevHU? fpt=ed25519-sha-256&cost=131072'},
    'public_keys': ['5RaeQAZEEQMFCLBEDrKtk5pLue4SuVkdw9Xb55
    ucpvQ6']}],
'version': '2.0'}
```

and finally, send it to the connected BigchainDB node:

```
>>>sent_transfer_tx=bdb.transactions.
    send_commit(fulfilled_transfer_tx)
>>>sent_transfer_tx==fulfilled_transfer_tx
```

True
   Bob is the new owner:

```
In [25]: fulfilled_transfer_tx['outputs'][0]['public_keys']
    [0]==bob.public_key
Out[25]:True
```

Alice is the former owner:

```
In [26]: fulfilled_transfer_tx['inputs'][0]['owners_before']
    [0]==alice.public_key
Out[26]:True
```

Note:

Obtaining asset ids:

You might have noticed that we considered Alice's case of consuming a CREATE transaction as a special case. In order to obtain the asset id of a CREATE transaction, we had to use the CREATE transaction's id:

```
transfer_asset_id=create_tx['id']
```

If you instead wanted to consume TRANSFER transactions (e.g., fulfilled_transfer_tx), you could obtain the asset id to transfer from the asset['id'] property:

```
transfer_asset_id=transfer_tx['asset']['id']
```

3.7 DECENTRALIZATION

Decentralization means that no one owns or controls everything, and there is no single point of failure.

Naturally, every hub in a BigchainDB system is claimed and constrained by an alternative individual or association. Even if the network lives within one organization, it's still preferable to have each node controlled by a different person or subdivision.

Utilize the expression "BigchainDB consortium" (or only "consortium") to allude to the arrangement of individuals or potentially associations who run the hubs of a BigchainDB network. A consortium requires some form of governance to make decisions such as membership and policies. The exact details of the governance process are determined by each consortium, but it can be much decentralized.

A consortium can increase its decentralization (and its resilience) by increasing its jurisdictional diversity, geographic diversity, and other kinds of diversity. There's no hub that has a long-term unique position in the BigchainDB network. All nodes run the same software and perform the same duties.

On the off chance that somebody has (or gets) administrator access to a hub, they can upset that hub (for example change or erase information put away on that hub), yet those progressions ought to stay secluded to that hub. The BigchainDB network can only be compromised if more than one third of the nodes get compromised. It's worth noting that not even the admin or superuser of a node can transfer assets. The only way to create a valid transfer transaction is to fulfill the current crypto-conditions on the asset, and the admin/superuser can't do that because the admin user doesn't have the necessary information (e.g., private keys).

3.7.1 IMMUTABLE

The word immutable signifies "perpetual after some time or unfit to be changed." For instance, the decimal digits of π are immutable (3.14159...). The blockchain community often describes blockchains as "immutable." If we interpret that word literally, it means that blockchain data is unchangeable or permanent, which is absurd. The data *can* be changed. For instance, a plague may drive humankind wiped out; the information would then get undermined after some time because of water harm, warm clamor, and the general increment of entropy.

The facts confirm that blockchain information is increasingly hard to change (or erase) than expected. It's more than simply "alter safe" (which infers purpose), blockchain information likewise opposes irregular changes that can occur with no aim, for example, information defilement on a hard drive. Therefore, in the context of blockchains, we interpret the word "immutable" to mean *practically* immutable, for all intents and purposes. (Language specialists would state that "immutable" is a term of workmanship in the blockchain network.)

The blockchain data can be made immutable in several ways:

1. **No APIs for changing or deleting data**: Blockchain software usually doesn't expose any APIs for changing or deleting the data stored in the blockchain. BigchainDB has no such APIs. This doesn't prevent changes or deletions from happening in *other* ways; it's just one line of defense.
2. **Replication**: All data is replicated (copied) to several different places. The higher the replication factor, the more troublesome it progresses toward becoming to change or erase all replicas.
3. **Internal watchdogs**: All nodes monitor all changes and if some unallowed change happens, then appropriate action can be taken.
4. **External watchdogs**: A consortium may opt to have trusted third-parties to monitor and audit their data, looking for irregularities. For a consortium with openly decipherable information, the public can go about as a reviewer.
5. **Economic incentives**: Some blockchain frameworks make it extravagant to change old stored information. Examples include proof-of-work and proof-of-stake systems. BigchainDB doesn't use explicit incentives like those data can be stored using fancy techniques, such as error-correction codes, to make some kinds of changes easier to undo.
6. **Cryptographic signatures**: These are frequently utilized as an approach to check if messages (for example transactions) have been altered en route, and as an approach to confirm who marked the messages. In BigchainDB, every transaction must be signed by at least one or more parties [13].
7. **Full or partial**: Partial reinforcements might be recorded occasionally, perhaps on attractive tape stockpiling, different blockchains, printouts, and so forth.
8. **Strong security**: Hub proprietors can receive and authorize solid security approaches [14,15].
9. **Node diversity**: Decent variety makes it so nobody thing (for example cataclysmic event or working framework bug) can bargain enough of the hubs.

3.8 FAULT TOLERANCE IN THE BIGCHAINDB SYSTEM

All the while protecting the versatility and trustless decentralization of both large-scale databases and decentralized blockchains is the principle target of the BigchainDB framework.

The following concepts are used while planning BigchainDB's safety efforts:

- **Benign deficiencies**: In the BigchainDB arrangement, hubs impart through a database which uses an issue tolerant accord convention, for example, Raft or Paxos. Subsequently we can expect that if there are 2f + 1 hubs, f benevolent flawed hubs can be endured (anytime), and every hub sees a similar request of keeps in touch with the database.
- **Byzantine deficiencies**: In request to work in a trustless system, BigchainDB fuses measures against noxious or flighty conduct of hubs in the 11 framework. These incorporate components for democratic upon exchange and square approval. The efforts to accomplish full Byzantine resilience are on the guide and will be tried with normal security reviews.
- **Sybil attack**: Deploying BigchainDB in a league with a high obstruction of section dependent on trust and notoriety disheartens the members from playing out an assault of the clones[]. The DNS framework, for instance, is living verification of an Internet scale appropriated league.

3.8.1   (BENIGN) FAULT TOLERANCE

One of the most common ways for a process to be faulty is for it to be unresponsive. That can happen, for example, if a hard drive fails or a CPU overheats. Such defects are known as benign faults or fail-stop faults. A consensus algorithm which enables a distributed system to come to consensus despite benign faults is said to be fault tolerant (FT). (It would be more precise to say "benign-fault tolerant," but it's not up to us.) In general, fault-tolerant consensus algorithms require at least 2f + 1 processes to be able to tolerate up to f faulty processes.

3.8.2   BYZANTINE FAULT TOLERANT

A Byzantine deficiency (likewise intuitive consistency, source congruency, blunder torrential slide, Byzantine understanding issue, Byzantine commanders issue, and Byzantine failure [16]) is a state of a PC framework, especially appropriated figuring frameworks, where parts may fall flat and there is defective data on whether a segment has fizzled. The term takes its name from a purposeful anecdote, the "Byzantine Generals Problem" [17], created to depict a circumstance wherein, so as to evade disastrous disappointment of the framework, the framework's entertainers must concur on a coordinated system, however a portion of these on-screen characters are inconsistent.

In a Byzantine issue, a segment, for example, a server can conflictingly seem both fizzled and working to disappointment recognition frameworks, showing various indications to various onlookers. It is hard for different parts to pronounce it fizzled and shut it out of the system, since they have to initially arrive at an agreement with respect to which segment has flopped in any case.

Byzantine adaptation to non-critical failure (BFT) is the steadfastness of a flaw tolerant PC framework to such conditions [18].

Byzantine shortcomings incorporate "ordinary" flaws, for example, hub crashes, yet additionally deviations from the convention, including particular non-support, prevarication, and other subjective conduct. For instance, a hub may appear collided with Joe, yet receptive to Mary, and moderate to-react to Albert. Or on the other hand a hub may very well drop a few exchanges, imagining they never occurred (which is one approach off convention). Or then again a hub may spam the system with duplicates of a similar exchange over and again. There are boundlessly numerous conceivable Byzantine shortcomings. They are, truly, any sort of flaw whatsoever.

In BigchainDB, regardless of whether up to 33% of the hubs "come up short" in any capacity, the remainder of the hubs can in any case come to understanding (agreement) on the following square. (I'm expecting all hubs have the equivalent "control.") There's no single purpose of disappointment and no single purpose of control. It's astounding that this is even conceivable. It's one of the marvels of current software engineering.

There are many! We had some bogus beginnings, yet in the end chose to utilize Tendermint.

Utilizing Tendermint methods there are a few contrasts in BigchainDB 2.0. For BigchainDB hub administrators, the most recognizable contrast is that there's never again a bunch wide MongoDB database (copy set). That is on the grounds that the hubs currently speak with one another utilizing Tendermint wire conventions. There's a neighborhood MongoDB database in each BigchainDB hub; however, they're all free. Upsetting the MongoDB on one hub won't influence any of the others. Another distinction is that all the replication, casting a ballot, and agreement rationale is finished by Tendermint, not MongoDB or BigchainDB.

What does BigchainDB programming do? It executes a state machine: a lot of states and state advances. The arrangement of state advances is the arrangement of substantial BigchainDB exchanges. There is BigchainDB code to develop substantial exchanges, and to check if a given exchange is valid. The BigchainDB state machine likewise incorporates MongoDB! MongoDB is utilized to store state and question state. You can think about the MongoDB database as giving BigchainDB an inherent "blockchain pilgrim."

A subjective JSON archive can be joined to each benefit, and to each move exchange. That usefulness can be utilized for things you wouldn't typically consider as "following resources."

To put it plainly, there has consistently been more to BigchainDB than code for agreement. The old BigchainDB accord code wasn't BFT, an issue that disturbed us for quite a while. We considered our very own considerable lot thoughts to make it BFT; however, we never discovered one that we could actualize in a sensible measure of time. Since we have Tendermint doing agreement, we can rest better realizing that this piece of the BigchainDB stack is strong.

Tendermint Core is a blockchain application stage; it gives what might be compared to a web-server, database, and supporting libraries for blockchain applications written in any programming language. Like a web-server serving web applications, Tendermint serves blockchain applications.

All the more officially, Tendermint Core performs Byzantine Fault Tolerant (BFT) state machine replication (SMR) for discretionary deterministic, limited state machines.

Tendermint is programming for safely and reliably duplicating an application on numerous machines. By safely, we imply that Tendermint works regardless of whether up to one- third of machines bomb in self-assertive ways. By reliably, we imply that each non-broken machine sees a similar exchange log and registers a similar state. Secure and reliable replication is a key issue in circulated frameworks; it assumes a basic job in the adaptation to non-critical failure of a wide scope of utilizations, from monetary forms, to decisions, to foundation organization, and past.

The capacity to endure machines bombing in self-assertive ways, including getting to be pernicious, is known as Byzantine adaptation to internal failure (BFT). The hypothesis of BFT is decades old; however, programming usage have just ended up mainstream as of late, due generally to the achievement of "blockchain innovation" like Bitcoin and Ethereum. Blockchain innovation is only a reformalization of BFT in an increasingly present day setting, with accentuation on distributed systems administration and cryptographic confirmation. The name gets from the manner in which exchanges are clustered in squares, where each square contains a cryptographic hash of the past one, framing a chain. By and by, the blockchain information structure really upgrades BFT plan.

Tendermint comprises of two boss specialized segments: a blockchain accord motor and a conventional application interface. The accord motor, called Tendermint Core, guarantees that similar exchanges are recorded on each machine in a similar request. The application interface, called the Application BlockChain Interface (ABCI), empowers the exchanges to be handled in any programming language. Not at all like other blockchain and agreement arrangements, which come pre-bundled with inherent state machines (like an extravagant key-esteem store, or a particular scripting language), engineers can utilize Tendermint for BFT state machine replication of uses written in whatever programming language and improvement condition is directly for them.

Tendermint is intended to be anything but difficult to utilize, easy to see, exceptionally performant, and helpful for a wide assortment of circulated applications.

Tendermint is comprehensively like two classes of programming. The top notch is composed of circulated key-esteem stores, as Zookeeper, etcd, and emissary, which use non-BFT agreement. The below average is known as "blockchain innovation," and comprises both digital forms of money like Bitcoin and Ethereum, and option dispersed record structures like Hyperledger's Burrow (Figure 3.6).

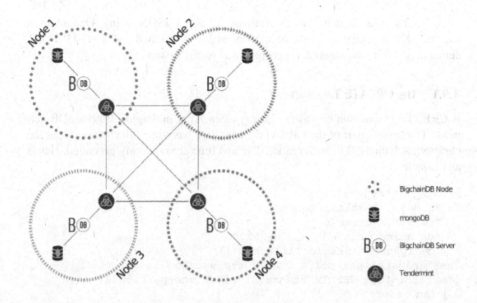

A four-node BigchainDB 2.0 cluster

FIGURE 3.6   Four node BigchainDB 2.0 cluster.

3.9   QUERYING BIGCHAINDB

A hub administrator can utilize the full intensity of MongoDB's question motor to look and inquiry all put away information, including all exchanges, resources, and metadata. The hub administrator can choose for themselves the amount of that inquiry control they open to outer clients. I don't get that's meaning?

Assume that Sergio Tillenham plans in vogue custom vehicles. He needed to make a reliable record of Sergio Tillenham vehicle proprietors, with the goal that potential purchasers could look into the past proprietors of every vehicle. (The possession history, or provenance, influences the estimation of a vehicle.) Moreover, in the event that somebody guarantees a vehicle was planned by Sergio Tillenham; however, it's not in the database, at that point it must be a phony. Likewise, if the vehicle is in the database yet the underlying record wasn't made and marked (cryptographically) by Sergio, at that point it's additionally phony.

Sergio chose to utilize BigchainDB. He made an arrangement with the top of the line vendors who sell his autos: in the event that a seller needs to sell his vehicles, at that point they should run a hub in his vehicle proprietorship database (fueled by BigchainDB). This gives every seller the additional advantage of having the full intensity of MongoDB to inquiry the database [7].

A BigchainDB exchange is a serialized article: a JSON string. One sends a BigchainDB exchange to a BigchainDB arranges in the body of a HTTP POST demand, and if it's substantial, it gets put away by the system.

3.9.1 THE CREATE TRANSACTION

A CREATE transaction begins the history of each car in Sergio's BigchainDB network. The "asset" part of the CREATE transaction has some information about the car itself. A name, color, and creation date and time are randomly generated. Here's an example:

```
{"data": {"type": "car",
   "name": "Restless Bonus",
   "color": "cream",
   "datetime_created": 1075128200,
   "designer": "Sergio Tillenham"}}
The"metadata" part of a CREATE transaction is set to:
{"notes": "The CREATE transaction for one particular car
   (an asset)."}
```

You may have noticed the strange format of the "datetime_created" in the above example, i.e., 1075128200. That's a POSIX time stamp, also known as a Unix time stamp. We stored times that way to make certain queries easier. (MongoDB also has Date objects, but BigchainDB doesn't support those yet.) As it happens, 1075128200 is the POSIX time stamp of 14:43:20 on January 26, 2004 (UTC). (We told the script to generate a random time between 30 years ago and 3 years ago.)

Each CREATE transaction is signed by Sergio Tillenham's private key and he is the first owner, i.e., the condition on the transaction's sole output says that it can only be transferred if Sergio Tillenham signs the TRANSFER transaction.

3.9.2 THE FIRST TRANSFER TRANSACTION

The first TRANSFER transaction (for each car) is special because it always transfers the car from Sergio to its first owner. We assumed that the first transfer happens within two years of the time the car was created.

The "asset" part of the first TRANSFER transaction looks like:

```
{"id":
"816c4dd7ae10b59a10f7016aacd685a5a41b402b3394a4264583d25
   851af1629"}
```

where the long hex string is the asset id of the car (and also the transaction ID of the CREATE transaction where it was created).

The "metadata" part of the first TRANSFER Transaction looks like:

```
{"notes": "The first transfer, from Sergio to the first owner.",
"new_owner": "Matthew Wheeler",
"transfer_time": 1101566600}
```

The name of the new owner is randomly generated. The transfer time is between zero and two years after the creation time, also randomly generated. In this case, 1101566600 is 14:43:20 on November 27, 2004 (UTC).

3.9.3 ALL OTHER TRANSFERS

The transfer to the *next* owner, if it happens at all, is assumed to happen within 365 days (about ten years). The time interval is randomly generated. More transfers are generated until the generated transfer time is in the future and therefore, hasn't happened yet.

The "asset" part of the first TRANSFER transaction looks the same as in the first TRANSFER transaction. The "metadata" part looks like Figure 3.7.

```
{"notes": null,
"new_owner": "Laurie Jones",
"transfer_time": 1228920200}
```

3.10   EXPLORING THE STORED DATA WITH MONGODB

Now that we've got a bunch of data stored in the MongoDB database, we can use MongoDB queries to answer some questions, such as:

- How many cream-colored cars did Sergio design?
- How many transfers happened in 2010?
- What was the longest car name?
- What's the average time that an owner owns a car?
- What's the maximum number of times a car has been transferred?

Before we can answer those questions, we need to get a tool that can connect to, and explore, a MongoDB database. There are many options, including:

- Graphical MongoDB apps such as MongoDB Compass, Studio 3T, Mongo Management Studio, NoSQLBooster for MongoDB, and Dr. Mongo.

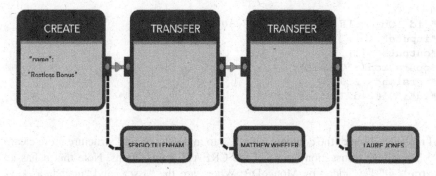

FIGURE 3.7   Shows the model for transfer method.

- The Mongo Shell
- One of the MongoDB drivers, such as official drivers for C, C++, C#, Java, Node.js, Perl, PHP, Python, Motor (Python async), Ruby and Scala, and community drivers for many other languages, including Go and Smalltalk.

The Mongo Shell (mongo) is free, and we can use it to start exploring:

```
$ mongo
MongoDB shell version v3.6.5
connecting to: mongodb://127.0.0.1:27017
MongoDB server version: 3.6.4
...
> show dbs
admin 0.000GB
bigchain 0.000GB
config 0.000GB
local 0.000GB
> use bigchain
switched to dbbigchain
> show collections
assets
blocks
metadata
pre_commit
transactions
utxos
validators
```

We see that there are several collections in the bigchain database. The main ones are transactions, assets, and metadata. We can get a look at an example document in the transactions collection using:

```
>db.transactions.findOne()
```

Here's one such document:

```
{
"_id":ObjectId("5b17b9fa6ce88300067b6804"),
"inputs":[...],
"outputs":[...],
"operation":"CREATE",
"version":"2.0",
"id":"816c4dd7...851af1629"
}
```

(I replaced some of the contents with "..." to make the overall structure more clear.)

The above transaction is one of the CREATE transactions. Note that it has an extra "_id" key added by MongoDB. Where are the "asset" and "metadata" keys and values? Did MongoDB forget them? No. When BigchainDB stores a CREATE transaction in MongoDB, it removes those and stores them in separate collections named assets and metadata.

TRANSFER transactions are stored slightly differently. We can find one using:

```
>db.transactions.findOne({"operation":"TRANSFER"})
Can you spot the difference in this TRANSFER transaction?

{
"_id":ObjectId("5b17b9fa6ce88300067b6807"),
"inputs":[…],
"outputs":[…],
"operation":"TRANSFER",
"asset":{
"id":"816c4dd7ae…51af1629"
   },
"version":"2.0",
"id":"985ee697d…a3296b9"
}
```

Answer: the "asset" key and value were not removed. Why?

A TRANSFER transaction doesn't define a new asset, it just points to an already existing asset. Its "asset" field doesn't contain any new information about assets, so it's not removed and stored in the assets collection.

Let's look at a random example document in the assets collection:

```
{
"_id":ObjectId("5b17b9fe6ce88300067b6823"),
"data":{
"type":"car",
"name":"Long Wave",
"color":"lilac",
"datetime_created":1093255577,
"designer":"SergioTillenham"
   },
"id":"96002ef8740…45869959d8"
}
```

We see that it contains two extra keys in besides the usual "data" key: the "_id" added by MongoDB and the "id" of the transaction it came from.

Below is an example document in the metadata collection; the story there is the same.

```
{
"_id":ObjectId("5b17ba006ce88300067b683d"),
"metadata":{
"notes":"The first transfer, from Sergio to the first owner.",
"new_owner":"Meagan Bowers",
"transfer_time":1058568256
   },
"id":"53cba620e…ae9fdee0"
}
```

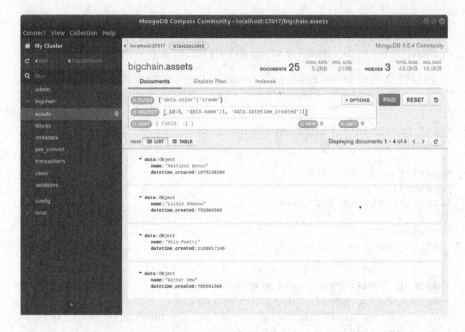

FIGURE 3.8   Shows a sample for MongoDB Compass.

We could do some queries using the Mongo Shell, but let's try one of the MongoDB apps with a nice graphical user interface: MongoDB Compass. When you connect, just use the default values (localhost:27017).

We can now answer some of the questions asked earlier. For sample, "How many cream-colored cars did Sergio design?" (Figure 3.8)

There are four results, i.e., Sergio designed four cream-colored cars. Note how we projected the results to suppress the _id, and to show the data.name and data. datetime_created.

We could also answer that question using the Mongo Shell:

```
>db.assets.find({"data.color":"cream"}).count()
4
```

We can also answer, "How many transfers happened in 2010?" The first step is to get the start time and end time as POSIX time stamps.

- 00:00:00 on January 1, 2010: 1262304000
- 23:59:59 on December 31, 2010: 1293839999

Now we can do the query (Figure 3.9):

The data $gte means "greater than or equal" and $lte means "less than or equal." They are two of the MongoDB's comparison query operators. The query means "find all metadata documents where metadata.transfer_time is greater than or equal to 1262304000 and less than or equal to 1293839999."

We see that there were four car transfers in 2010.

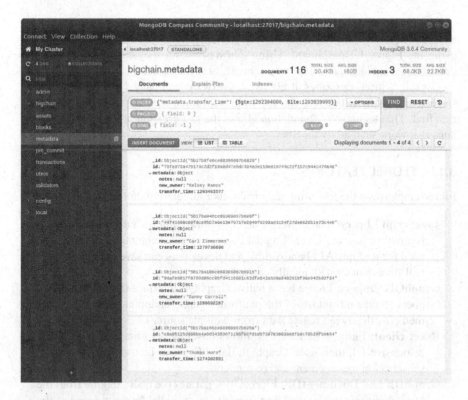

FIGURE 3.9 Shows a sample for MongoDB comparison for query operators.

3.11 ABOUT MONGODB

MongoDB is the main current, universally useful database stage, intended to release the intensity of programming and information for engineers and the applications they assemble. MongoDB is an open-source report database and driving NoSQL database. MongoDB is written in C++. The remarkable MongoDB engineering consolidates the best of both social and non-social databases, tending to the necessities of associations for execution, adaptability, adaptability and unwavering quality while keeping up the qualities of heritage databases. Download MongoDB Enterprise and Ops Manager and look at MongoDB Atlas for the most ideal approaches to run MongoDB.

3.12 FLUREE

Fluree, found online at Flur.ee, is a blockchain-based database. The company's flagship product, Fluree Advocate, empowers employees and customers to promote brand messages across their social media connections [19].

The company's software is built on the FlureeDB platform, which builds transactions into chained blocks, with each block representing a moment in time. FlureeDB is a blockchain technology, and apps like *Fluree* Advocate are decentralized apps.

Fluree also has a query engine that allows every query to yield a snapshot of data as it related to any block or point in time. Meanwhile, users can visualize their data using the Fluree UI dashboard. Using that dashboard, they can interact with the database, view blocks, and explore a schema, query, and stats.

Overall, Fluree describes itself as "a software company built upside down." Instead of starting with a specific app in mind, Fluree built the technology foundation first. That technology foundation allows the fast creation of multiple enterprise apps that work seamlessly together.

3.13   FLUREE FEATURES

Fluree emphasizes the following features in its FlureeDB platform:

JavaScript library: The Fluree JavaScript library allows a mini FlureeDB to run within your apps, enabling data subscriptions automatically. There's no need for multiple API bandwidths, and developers can save data by serving real-time queries with locally cached data.

GraphQL support: Fluree has a native GraphQL interface that allows developers to take advantage of the platform's single-endpoint query language, then effortlessly aggregate data from multiple sources.

React client: Fluree's React client allows developers to build a flexible and responsive UI, then fetch GraphQL data. This can be used for automated subscription management and re-renders.

GraphDB and DocumentDB: FlureeDB separates the query engine from the blockchain transactor, enabling support for multiple database types at the same time. This allows for horizontally scaling queries with low latency.

Time-based blocks: FlureeDB builds transactions into chained blocks, with each block representing a moment in time. Then, developers can use that query engine to represent any block at a particular moment in time.

Supports multiple consensus rules: The FlureeDb blockchain supports a variety of blockchain consensus mechanisms. Networks that need fast transactions might have low transaction needs (say, for internal transactions), while networks with high consensus needs can use more strict verification levels. Fluree has three permission levels (private, federated cluster, and full decentralization) to help businesses choose between a private or public blockchain—or somewhere in between.

Ultimately, the multi-consensus mechanism in Fluree allows businesses to create an internal, permissioned database for complete privacy, or create a *fully decentralized database* for complete transparency. They also have a "federated cluster" structure that's somewhere in between.

3.13.1   FLUREE APPS

Fluree has a number of different apps built on its blockchain platform. These apps are all decentralized and include Fluree Advocate and Fluree Cap Table. It also includes the Fluree App Factory.

3.13.1.1 Fluree Advocate

Fluree Advocate is described as the company's flagship app. It's an employee advocacy app marketed as a way to attract top talent to your organization. Social endeavors are 58% bound to draw in top ability and 20% bound to hold them. Advocate allows employees to fill positions faster, with employee referrals filled in half the time compared to those sourced from traditional career websites.

Advocate also encourages customer advocacy to increase conversions. As the Fluree Advocate sales page explains, "content shared from advocates are 7x more likely to convert as content shared directly by the company."

Overall, Fluree encourages the creation of better relationships, and customers sourced through advocates have 37% higher renewal rates. The term "advocate" is used in a similar way to "affiliate," although advocate appears to suggest a higher level of personal connection.

3.13.1.2 Fluree Cap Table

Fluree's Cap Table helps businesses issue stock and manage their equity in one place without getting bogged down in spreadsheets and paperwork. It aims to provide access to shareholders and employees, letting them transparently *monitor their investments*, adjust their vesting schedules, or access convertible instruments.

3.13.1.3 Fluree App Factory

Want to build apps on the FlureeDB blockchain yourself? The company has released its Fluree App Factory, which helps developers construct decentralized apps on the platform. Using the Fluree App Factory, you can leverage the FlureeDB blockchain database and a serverless, decentralized app platform to create the next era of enterprise apps.

REFERENCES

1. Drescher, D. *Blockchain Basics: A Non-Technical Introduction in 25 Steps*, Apress, 2017.
2. Nagasubramanian, G., Sakthivel, R. K., Patan, R., Gandomi, A. H., Sankayya, M. and Balusamy, B. Securing e-health records using keyless signature infrastructure blockchain technology in the cloud. *Neural Computing and Applications*, 2018, 1–9.
3. Karthikeyan, S., Patan, R. and Balamurugan, B. Enhancement of security in the Internet of Things (IoT) by Using X. 509 Authentication Mechanism. In *Recent Trends in Communication, Computing, and Electronics* (pp. 217–225). Springer, Singapore, 2019.
4. Rana, T., Shankar, A., Sultan, M. K., Patan, R. and Balusamy, B. An intelligent approach for UAV and Drone Privacy Security Using Blockchain Methodology. In *2019 9th International Conference on Cloud Computing, Data Science & Engineering (Confluence)* (pp. 162–167). IEEE, 2019.
5. Kumar, S. R. and Gayathri, N. Trust based data transmission mechanism in MANET using sOLSR. In *Annual Convention of the Computer Society of India* (pp. 169–180). Springer, Singapore, 2016.
6. Swan, M. *Blockchain: Blueprint for a New Economy*, 1st ed., O'Reilly Media, 2015.
7. Andreas M. Antonopoulos Mastering Bitcoin: Programming the Open Blockchain, 2nd ed., O'Reilly Media, 2017.
8. BigchainDBContributors, BigchainDB Documentation Release 0.5.0, July 04, 2016.

9. Bigchain DB 2.0 white paper, www.bigchaindb.com/whitepaper/, May 2018.
10. Muthuramalingam, S., Bharathi, A., Gayathri, N., Sathiyaraj, R. and Balamurugan, B. IoT Based Intelligent Transportation System (IoT-ITS) for global perspective: A case study. In *Internet of Things and Big Data Analytics for Smart Generation* (pp. 279–300). Springer, Cham, 2019.
11. QuillHashTeam, Understand and Build on BigchainDB (Series Part 1), December 21, 2018.
12. Armstrong, S. What is BigchainDB? https://opensource.com/article/18/12/Bigchain DB, December 28, 2018.
13. Namasudra, S., Devi, D., Choudhary, S., Patan, R. and Kallam, S. Security, privacy, trust, and anonymity. *Advances of DNA Computing in Cryptography*, (pp. 138–150), Chapman and Hall/CRC, 2018.
14. Shankar, A., Jaisankar, N., Khan, M. S., Patan, R. and Balamurugan, B. Hybrid model for security-aware cluster head selection in wireless sensor networks. *IET Wireless Sensor Systems*, 2018, 9(2), 68–76.
15. Karthikeyan, S., Rizwan, P. and Balamurugan, B. Taxonomy of security attacks in DNA computing. In *Advances of DNA Computing in Cryptography* (pp. 118–135). Chapman and Hall/CRC, 2018.
16. Bashir, I., *Mastering Blockchain: Distributed Ledger Technology, Decentralization, and Smart Contracts"*, 2nd ed., Packt, Mumbai, 2018.
17. McConaghy, T. and Marques, R. BigchainDB: A Scalable Blockchain Database, June 8, 2016, ascribe GmbH, Berlin, Germany.
18. Tanner, J. Challenges to Overcome while working with BigchainDB, https://blog.indorse.io/4-challenges-overcome-while-working-with-bigchaindb-5f8d50a880a4, December 18, 2017.
19. blockchain oodles, Use Cases For BigchainDB In Business, https://dzone.com/articles/a-growing-list-of-prominent-use-cases-for-bigchain, July 29, 2018.

4 Blockchain Databases 2

Anurag Pandey, Abhishek Kumar, Achintya Singha, N. Gayathri, and S. Rakesh Kumar

CONTENTS

4.1 INTRODUCTION

This chapter deals with the concept of databases in blockchain technologies and its scope in current trend and scenarios. In the present time, blockchain database is a conveyed database that maintains a persistently growing tamper-proof data structure blocks which hold batches of individual transactions, without any involvement of third parties. Due to which the demand for blockchain and related technologies is going to increase more in the coming days. Here, many concepts are discussed thoroughly like BigchainDB, file storage in BigchainDB, smart contracts, transaction concepts and permissions, privacy in blockchain databases and many others with its future aspects.

4.2   BIGCHAINDB: THE BLOCKCHAIN DATABASE

"A database is an organized way of storing a bunch of information, or data, that are stored in electronic in a computer system. A database is usually managed by a database management system (DBMS). Taking as together, the data and the DBMS, along with the applications that are associated with them, are referred to as a database system, often abbreviated to just database" [1,19]. Similarly, in terms of blockchain technology, blockchain database consists of many decentralized nodes. Every node takes an interest in the organization all the nodes to check the new increments to the blockchain and is equipped for entering new data into the database. For expansion to be made to the blockchain, most of the nodes must arrive at a consensus. This consensus system ensures the security of the system [9,10].

There are many databases available to store our information, namely:

- traditional database
- decentralized database
- centralized database
- relational database

In a centralized database, the input information from the users is stored at a fixed position such that different users from different locations can access it. This type of database helps the users to access the database from remote areas.

In a decentralized database, rather than storing data at a fixed central point of control, data are stored in a distributed architecture at different geographical locations. This implies that it is a group of independent database instances which have no direct logical connections established between them. This is one of the most exciting properties of blockchain technology.

In a relational database, information's are distributed among a set of tables where information are stored in a pre-structured table. The table consists of a number of rows and columns, where rows contain instance for the data according to its type and columns contains an entry for the information for a specific purpose.

The idea of blockchain has become popular due to massive-scale adoption of its different achievements like a smart contract (to be discussed in the following topic), P2P digital currencies called cryptocurrencies, under the name of bitcoin which is a revolutionary in the digital market [11]. BigchainDB is one of the best examples of blockchain-based technology, that integrates the concept of blockchain and traditional database, to provide best in class output that utilizes the areas of both the worlds.

BigchainDB is an application that has properties similar to the blockchain (e.g., decentralization concept, immutability property, owner-controlled assets, etc.) and the database properties (e.g., high transaction rate, low latency, indexing, querying processing of structured data, etc.) [6]. It is an open-source application, whose first version is released in February 2016. Its latest edition is 2.0 which has significant improvement over the previous one and is continuously upgrading. BigchainDB enables engineers and ventures to send blockchain evidence of ideas, different stages, and applications with a versatile blockchain database [6]. BigchainDB has various applications in different areas like supply chain, Internet of Things (IoT), stock market, financial system, etc.

BigchainDB is preferred due to its high throughput, high latency, high capacity, and low communication cost.

4.3   SMART CONTRACT: AN OVERVIEW

In the last few years, blockchain technologies become an emerging area due to its anonymous behavior, P2P transaction, distributed consent, and the most important are its decentralized immutable property. Due to which many problems and technical glitches related to the technology have been resolved.

This segment focused on a popular computer program used in blockchain technologies integrated with significant applications of blockchain technologies.

4.3.1   What Is Smart Contract?

The word "contract" is defined as a legal and formal agreement that is undersigned by two parties. Similarly, smart contract is also a regular contract that exits in the form of code in a computer program having its own verifying, executing, and tamper-resistant properties (Figure 4.1).

Nick Szabo introduced the concept of smart contract in 1994. He is a computer researcher, legitimate scholar, and cryptographer known for his investigation in computerized deals and digitized cash. Nick Szabo defined smart contract in the form of "A computerized transaction protocol that executes a term of a contract" [2]. He suggested translating contractual clauses into code and placing them into a property that can self-execute them. However, in blockchain technologies, the meaning of smart contracts has been coined. Within the blockchain terminologies, smart contracts are coded scripts recorded on the blockchain.

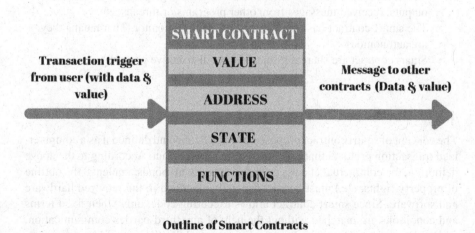

Outline of Smart Contracts

FIGURE 4.1   Outline of smart contract.

A smart contract allows executing code, only when fixed terms and conditions are matched without the help of any third parties. These are legal contracts which are purely written in a digital form of code of the agreement, which is based on blockchain. These self-executing contracts come into existence when terms and conditions met by parties involved are satisfied. Then only these contracts will execute it, causing the exchange of assets between the parties. These assets may be money (a cryptocurrency like Bitcoin), property, content or other like values between the users.

A smart contract may consist of many properties like different values, addresses, functions, and states. It always accepts the transaction as an input executes the relating code and furthermore, triggers the output result. After the complete process depending upon the logic implemented, state changes. The procedures are trackable and irreversible. Since 2008 when blockchain innovation appears through Bitcoin cryptocurrency, the significance of smart contract integration of blockchain technology becomes a key area to develop. Because, it gives peer to peer (P2P) transaction and database, that can be maintained openly in a protected manner in a trustful situation.

For the implementation of smart contract in blockchain technologies programming languages like Solidity has been preferred. It is a high-level programming language used to implement smart contracts. Solidity is used to execute the smart contract in different developing blockchain platforms like Ethereum, ErisDB, Zeppelin, and many more.

Some of the critical characteristics of a smart contract are:

- The smart contract is machine executable codes runs on different platforms of blockchain technologies.
- The smart contract is part of a single application program. That means each program is created and runs on a single application.
- The smart contract is event-driven programming. That means the flow of the program is determined by different events like user actions (key pressed on the keyboard, mouse clicks, voice command, fingerprint scanner), sensor outputs, received messages from other programs or threads, etc.
- The smart contract has once created no need to monitor. That means they are autonomous.
- Smart contract are shareable; means have distributive property.

4.3.2 How Smart Contract Works?

The concept of smart contract proposed by Nick Szabo and defined it as a computerized transaction protocol that executes a term of a contract. According to the above definition, the contractual clauses (security agreement, bonds, contents, the outline of property rights, etc.) must be encoded and embedded in the required hardware and software. Since smart contract allows executing code, only when fixed terms and conditions are matched without the help of any third-parties communication, which also secures the system from malicious attack. In the case of blockchain terminologies, smart contracts are coded scripts placed on the blockchain, which has the executive power. One can activate the transaction process to a smart contract by using its own or unique allotted address by the blockchain technologies.

Let us take an example to get a better understanding of the above scenario on working of smart contracts. Suppose you want to rent or sell your vehicle to someone in the blockchain network, so you can quickly deploy a smart contract in the above existing system. Different features regarding the car are stored in the blockchain, and those who are belonging to the same network can show their interest but cannot allow manipulating the existing features. In this way, we can find a buyer for the vehicle without involving any third party for the same.

There are wide ranges of dynamic applications; blockchain-based smart contracts could offer several significant advantages:

- High speed and quick real-time updates
- Privacy
- High accuracy
- Lower execution risk [7]
- Fever intermediaries
- Minimum cost

4.3.3 Use Cases of Smart Contracts

4.3.3.1 Supply Chain

The supply chain system consists of different stages of the transaction. Each step consists of some terms and related conditions. The various areas of supply chain management system are like the food packaging sector, shipment sector, power supply system, and many more. Since multiple system layers are involved in this area. So it is crucial to make it as a reliable and trustful management system. This can be done using a smart contract integrated with blockchain technologies. Digital ledger database helps supply chain management system more transparent, reliable, trustful and most important without any third party interrogation (Figure 4.2).

4.3.3.2 Internet of Things

The Internet of Things (IoT) is an arrangement of interrelated gadgets, mechanical pieces of equipment, advanced machines, items, living bodies or even individuals that are furnished with its own unique identity and the capacity to move information over a system without expecting human-to-human or human-to-computer association. The IoT is one of the most optimistic areas when integrated with a smart contract concept. It helps IoT devices to emerge in a more autonomous way compared to past behavior. Some of the areas where blockchain technology already used are a smart home, smart city, smart grids, industrial internet, connected cars, and many more [13] (Figure 4.3).

4.3.3.3 Insurance Claim

Insurance is a formal agreement from financial loss. But, conventional insurance system takes long time process to claim on your policy. There is much vagueness

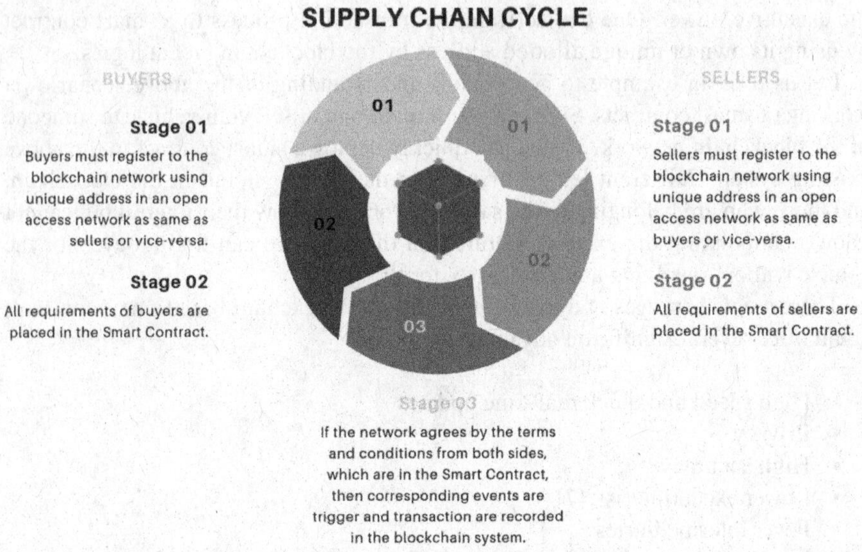

SUPPLY CHAIN CYCLE
BUYERS
SELLERS
01
01
02
02
03
Stage 01
Buyers must register to the blockchain network using unique address in an open access network as same as sellers or vice-versa.
Stage 02
All requirements of buyers are placed in the Smart Contract.
Stage 01
Sellers must register to the blockchain network using unique address in an open access network as same as buyers or vice-versa.
Stage 02
All requirements of sellers are placed in the Smart Contract.
Stage 03
If the network agrees by the terms and conditions from both sides, which are in the Smart Contract, then corresponding events are trigger and transaction are recorded in the blockchain system.

FIGURE 4.2 Outline of supply chain.

Embedded Mobile
Security
Travel & Tourism
Innovation
Internet Of Things (IoT)
Energy Use
Digitalization
Networking
Smart Homes & Cities

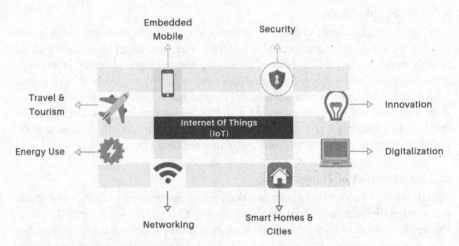

FIGURE 4.3 Outline of IoT framework.

emerges between various stakeholder during preparation. A smart contract-based system can streamline the procedure and utilizing blockchain technology [14], by which everything can be made straightforward just as system secure without third-party intercession into it (Figures 4.4 and 4.5).

There may more use cases of smart contacts like in the stock market, financial system [8], healthcare system [12], digital right management, real estate, and so on.

INSURANCE CLAIM CYCLE

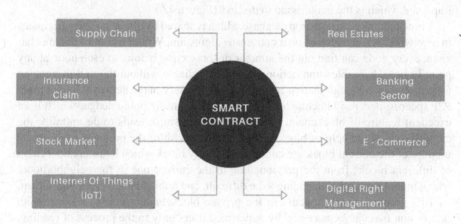

INSURANCE COMPANY

Stage 01

Different Insurance company must register to the blockchain network using unique address in an open access network as same as users or vice-versa.

Stage 02

All requirements of insurance companys are placed in the Smart Contract.

Smart Contract

1. Its types
2. Different Criteria
3. Terms & Conditions

USERS

Stage 01

Users must register to the blockchain network using unique address in an open access network as same as insurance company or vice-versa.

Stage 02

All requirements of users are placed in the Smart Contract.

Stage 03

Finally, if the network agrees by the terms and conditions from both sides, which are in the Smart Contract, then corresponding events are trigger and transaction are recorded in the blockchain system.

FIGURE 4.4 Outline of insurance claim.

Supply Chain

Insurance Claim

Stock Market

Internet Of Things (IoT)

SMART CONTRACT

Real Estates

Banking Sector

E - Commerce

Digital Right Management

FIGURE 4.5 Use case of smart contracts.

4.4 TRANSACTION CONCEPTS AND PERMISSIONS

The popularity of e-commerce is multiplying in our day-to-day life, but what is lacking in it is the credibility of transaction. As all the related data are stored at a single trading platform, it is easy to manage or control the services for the central bodies, and it is also easy to breach the security for the criminals. Therefore, only trading in all has permissions to give related data or information. If criminals attack on trading center then it is easy to catch them, without interfering the whole transaction system. Government is unable to maintain the statistics due to privacy on e-commerce. By adopting any algorithm of blockchain, it will improve the privacy of transaction

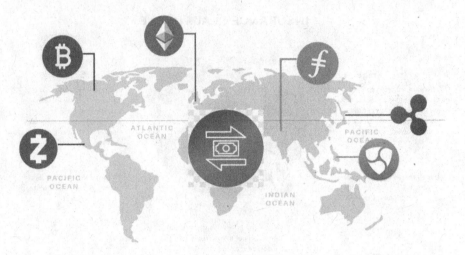

FIGURE 4.6 Transaction concepts and permissions.

and worsen the overall performance. For better performance, blockchain must be improved which is the major issue to discuss (Figure 4.6).

A blockchain is a transaction database which is shared by all the nodes participates in a network, which is based on a consensus algorithm. With the help of this mechanism, every node can find out the number of transaction belong to each node at any point in time. It enables transactions which are reliable without the tool to imagine is centrally, even though there is an unreliable adversary in the network. By adopting P2P (peer-to-peer) architecture, each peer stores an utterly public ledger which is an excellent feature of blockchain. Using this mechanism, it leads to decentralize the credibility with the help of blockchain. Overall, a hash of the previous block and a mixture of the current block are contained in every block which helps create a chain of different blocks from the previous one to the current one in the neighborhood, which makes the double-spending to be difficult, and public data to every participant.

This problem usually occurs in the private blockchain which is a ledger with permission that can be accessed by authorized users only in the process of reading, writing, and consensus. Not only the transaction of the token but also the transaction of the assets among corporations occurs in the private blockchain.

Each transaction must be processed immediately, and large transactions coincide. So the Bitcoin-derived blockchain can hold distributed ledgers with the great promises, but it cannot improve the performance with the help of this algorithm on e-commerce. It can solve the lack of credibility of transactions, but it does not help in better performance on throughput, capacity, and latency.

4.4.1 Permissioned Blockchain Technology

Bitcoin-derived blockchain implicitly defined and implemented the Nakamoto consensus, which is the distributed ledgers. There are lots of hurdles which come in

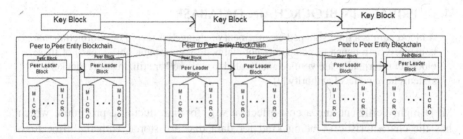

FIGURE 4.7 Typical permissioned blockchain architecture.

the way of Bitcoin-derived blockchain algorithm. There are lots of issues related to throughput, latency, capacity and communication cost in the blockchain algorithm.

The four types of blocks [5] mainly used in this technology are:

- peer block
- key block
- peer leader block
- micro block

Every neighborhood has a header that contains the reference of its predecessor uniquely defined, mainly cryptography hash of predecessor leader. Peer leader block includes a hash value of multiple micro blocks, key block contains multiple peer blocks, peer block contains peer leader blocks and multiple micro blocks, and micro block contains specific transactions that occur in one peer (Figure 4.7).

Permissioned block technology contains three parts mainly:

- Peer Inner Blockchain Protocol (PIBP)
- Permissioned Trusted Trading Network Consensus Algorithm (PTTNCA)
- Permissioned Trusted Blockchain Architecture (PTBA)

These technologies are widely used in blockchain technology. Lots of work has been done to improve throughput, latency, capacity, and communication cost.

4.4.2 PERMISSIONLESS BLOCKCHAIN TECHNOLOGY

A blockchain-distributed ledger records all transactions that take place on the network, which is a heart of blockchain technology. Using cryptography, which guarantees that once the operation has been added, it cannot be changed further, as a data store in the blockchain database is appended after the fact. This is the property of immutability.

The first and most widely used application of blockchain is the Bitcoin cryptocurrency, by adding the same characteristics as Bitcoin with adding a platform for a distributed implementation. Both Bitcoin and Ethereum are public permissionless blockchain technology class public network which is open to all participants interact anonymously with each other.

4.5   PRIVACY IN BLOCKCHAIN DATABASE

A blockchain is a time-stamped arrangement of changeless record of information that is mapped by a bunch of computers not claimed by any single object, each of these blocks of information are verified and limited to one another utilizing rule. Blockchain systems have no focal authority [4].

"It is highly efficient and a cost-effective way for introducing a procedure which has the characteristic of being kind for data that are stored with real-time high safety. However, while concerning on security issues of networking and privacy of communication, that has been rapidly grown, security is vital and has become a top popular topic in the area of communication and networking [3]." This network regularly updated by spreadsheets you have basic knowledge of blockchain exit and share and counting by consult database the blockchain database store in any one single location its data is assessable by anyone the internet. The blockchain technology is encouraged for its implementation if the highly secure and privacy-preserving decentralized system where transactions are not under the control of any third-party organizations [15–17]. The blockchain innovation in which information is encoded and disseminated over the whole networks, to build the security of the information put away in the proposed system [16–18]. Every one of the clients' blocks is hashed, and limitless hashes of the exchanges are put away in the blockchain.

4.6   FILE STORAGE

Decentralizing file storage techniques on the internet brings us clear benefits in different areas. Distributing data throughout the networks protects files from getting hacked or lost by the others (Figure 4.8).

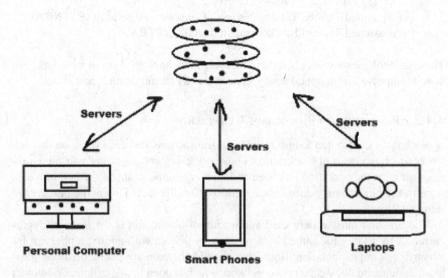

FIGURE 4.8   Client-server model.

A web made up of totally decentralized sites can accelerate record movement and streaming time. Such an improvement isn't just helpful; it's an essential move up to the webs as of now over-burden content-delivery systems.

4.6.1 CENTRALIZED VS. DECENTRALIZED STORAGE TECHNIQUES

In the centralized storage technique, the input information from the users is stored at a fixed place such that different users from different locations can access it easily. Because of this centralization at a place, all the data is put together at one spot. This makes an easy target spot for potential hackers to breach the data.

In the decentralized storage system, the data are not stored by a single user, as a single entity. Everyone in the network owns the data.

Peer-to-peer file storage emerged as an alternative to centralized clouds without the ideological and practical dangers of centralization.

All files are united within one swarm, where there is a common language in an integrated file system, and all peers are shared within the entire system, which allows people to discover and transfer files to each other.

4.6.2 DATA MANAGEMENT

Data management plays a crucial role in every technology to be successfully adopted. The users that use or provide data on different platforms such as social media platforms, platforms related to digital transactions, stock market, healthcare system, etc., must be managed properly. This chapter designs a storage technique to manage personal data based on blockchain and cloud storage platforms. The proposed storage and sharing plan doesn't rely upon any outsider, and no single gathering has the total capacity to influence the preparation.

4.7 CONCLUSIONS

This chapter discussed a broad aspect related to the area in the blockchain databases, and different tools and techniques used in the current stage of the technology. The concept of BigchainDB in the blockchain terminology makes the database more trustworthy and reliable as compared to the traditional methodologies, but still many areas are untouched. This chapter discussed the details about a different aspect of the smart contract. As the smart contract is a regular contract that exits in the form of code in a computer program having its own verifying, executing, and tamper-resistant properties. However, smart contracts and their development are still in the beginning stages. There is a much more broad perspective in this area. Still there are some issues that needs to be addressed like scalability, flexibility, privacy issue, as the programmed code is available publicly to everyone present in the blockchain network, which may not be favorable for some application. This chapter discussed how useful it could be with the integration of blockchain and smart contracts in the coming few years. This chapter also discussed how blockchain provides anonymous behavior, peer-to-peer transaction, distributed consent, and its decentralized, immutable property for the users. Finally, we discussed different use cases of smart contracts. Each of these applied areas shows how convenient it is if we use smart contract in the blockchain technologies. There are still more areas we are going to see in the future where we apply the integration of smart contract and blockchain. Transaction and privacy in the blockchain databases are the key features in the blockchain technology. This chapter explores their different functionality.

From the above content, it is clear that, as a research point of view, blockchain databases are still needed to be the focus in the coming days, as early development is the foundation of any technology.

REFERENCES

1. Pee, S. J., Kang, E. S., Song, J. G., & Jang, J. W. (2019). Blockchain-based smart energy trading platform using smart contract. *1st International Conference on Artificial Intelligence in Information and Communication*, ICAIIC 2019, 322–325. https://doi.org/10.1109/ICAIIC.2019.8668978
2. Yi, H. (2019). Securing e-voting based on blockchain in P2P network. *Eurasip Journal on Wireless Communications and Networking*, 2019(1), 1–9. https://doi.org/10.1186/s13638-019-1473-6
3. Rosic, A. (2019). What is Blockchain Technology? A Step-by-Step Guide For Beginners. Retrieved from https://blockgeeks.com/guides/what-is-blockchain-technology/
4. Min, X., Li, Q., Liu, L., & Cui, L. (2016). A Permissioned Blockchain Framework for supporting instant transaction and dynamic block size. *Proceedings—15th IEEE International Conference on Trust, Security and Privacy in Computing and Communications, 10th IEEE International Conference on Big Data Science and Engineering and 14th IEEE International Symposium on Parallel and Distributed Processing with Applications*, IEEE TrustCom/BigDataSE/ISPA 2016, 90–96. https://doi.org/10.1109/TrustCom.2016.0050
5. Becky M. H. (2018, October 22). BigchainDB. Retrieved from https://en.bitcoinwiki.org/wiki/BigchainDB.
6. Kumar, P., & Kumar, A. (2019). Information technology impact on E-commerce business growth. *International Journal of Innovative Technology and Exploring Engineering*, 23(5), 1014–1017.
7. Mohanta, B. K., Panda, S. S., & Jena, D. (2018). An overview of smart contract and use cases in blockchain technology. *2018 9th International Conference on Computing, Communication and Networking Technologies*, ICCCNT 2018, 1–4. https://doi.org/10.1109/ICCCNT.2018.8494045
8. Kumar, P., Kumar, D., & Kumar, A. (2019). Customers perception ATM services. *International Journal of Innovative Technology and Exploring Engineering*, (7), 2504–2506.
9. Laurence, T. (2017). *Blockchain for Dummies*. Hoboken, NJ: John Wiley & Sons.
10. Norton, J. (2016). *Blockchain: Easiest Ultimate Guide to Understand Blockchain*. s.l.: CreateSpace Independent Publishing Platform.
11. Gates, M. (2017). *Blockchain: Ultimate Guide to Understanding Blockchain, Bitcoin, Cryptocurrencies, Smart Contracts and the Future of Money*. s.l.: Wise Fox Publishing.
12. Nagasubramanian, G., Sakthivel, R. K., Patan, R., Gandomi, A. H., Sankayya, M., & Balusamy, B. (2020). Securing e-health records using keyless signature infrastructure blockchain technology in the cloud. *Neural Computing and Applications*, 32(3), 639–647. https://doi.org/10.1007/s00521-018-3915-1.

13. Muthuramalingam, S., Bharathi, A., Gayathri, N., Sathiyaraj, R., & Balamurugan, B. (2019). IoT based intelligent transportation system (IoT-ITS) for global perspective: A case study. *Internet of Things and Big Data Analytics for Smart Generation* (pp. 279–300). Springer, Cham.

14. Rana, T., Shankar, A., Sultan, M. K., Patan, R., & Balusamy, B. (2019). An intelligent approach for UAV and drone privacy security using blockchain methodology. *In 2019 9th International Conference on Cloud Computing, Data Science & Engineering (Confluence)* (pp. 162–167). IEEE.

15. Khari, M., Garg, A. K., Gandomi, A. H., Gupta, R., Patan, R., & Balusamy, B. (2019). Securing data in internet of things (IoT) using cryptography and steganography techniques. *IEEE Transactions on Systems, Man, and Cybernetics: Systems.*

16. Karthikeyan, S., Patan, R., & Balamurugan, B. (2019). Enhancement of security in the internet of things (IoT) by using X. 509 authentication mechanism. In *Recent Trends in Communication, Computing, and Electronics* (pp. 217–225). Springer, Singapore.

17. Shankar, A., Jaisankar, N., Khan, M. S., Patan, R., & Balamurugan, B. (2018). Hybrid model for security-aware cluster head selection in wireless sensor networks. *IET Wireless Sensor Systems*, 9(2), 68–76.

18. Rahman, M. A., Ali, J., Azad, S., & Kabir, M. N. (2017). A performance investigation on IoT enabled intra-vehicular wireless sensor networks. *International Journal of Automotive and Mechanical Engineering (IJAME)*, 14(1), 3970–3984.

19. What is a database? (n.d.). Retrieved from https://www.oracle.com/in/database/what-is-database.html

5 Blockchain Use Cases in Big Data

Harsh Kumar, Khushbu Agrawal, M. R. Manu,
R. Indrakumari, and B. Balamurugan

CONTENTS

5.1   INTRODUCTION

Blockchain in big data is considered as the most promising developing technologies in academics and industries as well. Blockchain technology is a way to evolve automation along with business process management [1]. Blockchain technology along with big data increases transparency and privacy of data. Data is stored in blocks that contain the hash of that particular block and hash of the previous block and the hash is executed using SHA256. On a large-scale environment, blockchain is a decentralized system (independent system) for storing and transferring information [2]. To increase the security and quality of data, blockchain uses secured cryptographic distributed database technology, thereby increasing the efficiency. Blockchain prevents data leakage and to make any changes in data it is necessary to get multiple permissions, thus hacking becomes difficult for cybercriminals. Blockchain has many applications in every field like medical record system, banking, social media, etc., in the field of medical record system it witnesses for utilizing safe and secure health care data management [3]. Space giants like the U.S. National Aeronautics and Space Administration (NASA) and the European Space Agency (ESA) are studying ways to employ blockchain for their mission to make it more powerful. Blockchain can be an efficient mode for online data storage. Blockchain along with big data provides a unique solution to how banks can improve the way they protect their customer's digital privacy. Addition to its uses, it also acts as a filter to catch financial fraud, it serves as an effective barrier to data theft.

    The collaboration of big data and blockchain improves the overall quality of the data being captured, and the quick transfer of the data which is causing quite a stir.

Blockchain will give businesses to confidently identify the integrity of the data being generated. Accelerated progress and major changes are being made daily when it comes to new blockchain technologies, ushering in new developments in the big data analytics field. As more people start using blockchain-driven services, analytics also must be capable of generating high-quality data insights from the existing data.

5.1.1 An Overview about Blockchain Management

Blockchain, the foundation of Bitcoin, has received extensive attention in the last two decades. The blockchain serves as a perpetual record that allows transactions to occur in a decentralized manner [4]. Blockchain-based applications are springing up, covering numerous fields including financial services, banking, social media, reputation systems, and the Internet of Things (IoT) [5], and so on. However, there are still many difficulties and challenges of blockchain technology such as scalability and security problems that are to be resolved.

Blockchain is a sequence of blocks that holds a complete record of transactions like traditional public entries. Figure 5.1 illustrates an example of blockchain technology. The blockchain contains a previous block hash in the block header, a block has only one parent block. It is worth noting that parent blocks (ancestors) hashes would also be stored in the blockchain. The first block of a blockchain, which has no parent block, is called genesis block [6].

A block consists of the block header and the block body, the block header indicates which set of block validation rules to follow, the hash value of all the transactions in the block, a 256-bit hash value that points to the previous block. The block body comprises a transaction counter and transactions. The maximum number of transactions that a block can contain depends on the size of the block and each transaction size. To validate the authentication of transactions, blockchain uses an asymmetric cryptography mechanism. A digital signature-based technology, i.e., asymmetric cryptography is used in an untrustworthy environment. Blockchain has shown its potential for transforming the traditional industry with its key characteristics: decentralization, persistence, anonymity, and audibility [7].

The key characteristics of blockchain are as follows:

- Decentralization: In conventional centralized transaction systems, every transaction is required to be validated through the central trusted agency (e.g., the central banks), necessarily resulting in the cost and the

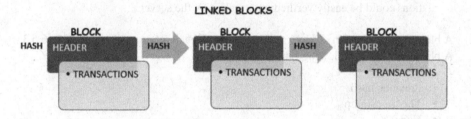

FIGURE 5.1 Blockchain technology.

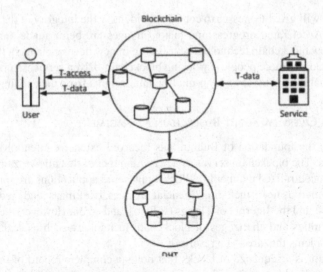

FIGURE 5.2   Decentralized system. (From Karafiloski, E. and Mishev, A., Blockchain solutions for big data challenges: A literature review, *IEEE EUROCON 2017—17th International Conference on Smart Technologies*, 2017.)

performance at the central servers as shown in Figure 5.2. In contradiction to the centralized model, the third party is no longer required in the blockchain. Consensus algorithms in blockchain are used to maintain data consistency in distributed networks worldwide.

- Persistency: The transactions can be validated instantly and invalid transactions would not be accepted by honest miners. It is almost impossible to delete or clear transactions once they are included in the blockchain. Blocks that contain invalid transactions can be discovered immediately and rejected instantly without and delay.
- Anonymity: Every user can interact with the blockchain with a generated unique address, which does not reveal the real identity of the user interacting. Note that blockchain cannot guarantee the perfect privacy protection due to the intrinsic constraint.
- Auditability: Blockchain stores transaction data about the user's balance based on the Unspent Transaction Output (UTXO) model. Once the current transaction is recorded into the blockchain, the once which are referred unspent transactions switches from unspent to spent. So that the transactions could be easily verified and tracked by the server.

A blockchain is a chain of blocks connected with each other as shown in Figure 5.3. A block consists of following four parts:

- Previous hash
- The timestamp
- Nonce
- Merkle tree root

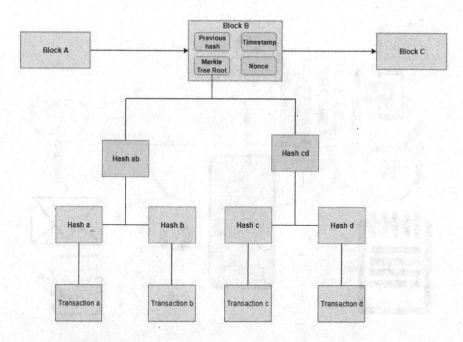

FIGURE 5.3   Block diagram of blockchain.

5.1.2 · BIG DATA TECHNOLOGY AND USES

Big data is a term defined for data sets that are large or complex in which traditional data processing applications are inadequate [8]. Big data comprises of analysis zing, capturing the data, data creation, searching, sharing, storage capacity, transfer, visualization, and querying and information privacy as shown in Figure 5.3. Big data are a more varied and complex structure with the difficulties of storing, analyzing and visualizing for further processes or results. The method of analysis of massive amounts of data to unveil hidden patterns and secret correlations is termed as big data analytics. Figure 5.4 shows big data technology. These useful pieces of information for companies or organizations is the goal of gaining richer and deeper insights and getting an advantage over the competition. This is the reason why big data implementations need to be analyzed and executed as accurately as possible.

Big data and its analysis are in the middle of modern science and business. These data are obtained from online transactions, emails, videos, audios, images, clickstreams, logs, posts, search queries, health records, social networking interactions, science data, sensors, and mobile phones and their applications. These data are stored in databases that expand massively and become difficult to capture, form, store, manage, share, analyze, and visualize via typical database software tools.

FIGURE 5.4   Big data technology.

5.2   USE OF BLOCKCHAIN IN BIG DATA

The advantage or features of the blockchain is that it has grown to become a dis-seminated, demoralized network of data entries. The data being stockpile or its coherence cannot be reign by anyone. Despite that, its records or entries are checks continuously by the various networking devices. Falsify data cannot enter the chain while storing on the one computer because it won't match the parallel data held by the other computers or machines. This is the reason for using the blockchain in big data. There are at least five peculiar ways blockchain data can help data scientists in general and blockchain technology in big data.

5.2.1   ENSURING TRUST (DATA INTEGRITY)

Data documented on the blockchain are realistic and sensible because they must have gone through a substantiation process which assures or safeguard its quality. It also provides for transparency, since activities and transactions that take place on the blockchain network can be imitated.

In 2018, Lenovo illustrates this use case of blockchain technology to disclose counterfeit documents and forms. The company used blockchain technology to legalize physical chronicle which were encoded with digital signatures. The digital signatures are processed by computers and the authenticity of the document is sub-stantiated through a blockchain record or entry.

Mostly, data probity is ensured when details of the origin and interactions concerning a data block are archived on the blockchain and automatically verified before it can be acted upon. Blockchain network is fortified by a number of computers called nodes, and these nodes confirm the transaction on this network. It is also secured by a cryptography algorithm, which is very tough to crack.

5.2.2 PREVENTING MALICIOUS ACTIVITIES

Since blockchain uses consensus algorithm and digital signatures to validate transactions, it is impossible for a single unit to pose a menace to the data network. A node (or unit) that begins to act abnormally can easily be analyzed and abolished from the network.

Because the network is so demoralized, it makes it almost impossible for a single party to provoke enough computational power to alter the verification criteria and allow undesirable data in the system. To amend the blockchain rules, a majority of nodes must be pooled together to create an accord. This will not be feasible for a single bad actor to achieve.

5.2.3 MAKING PREDICTIONS (PREDICTIVE ANALYSIS)

Blockchain data, just like other varieties of data, can be analyzed to divulge valuable intuition into the behaviors, trends and as such can be used to anticipate future outcomes. What is more, blockchain provides organized data accumulated from individuals or individual devices.

In predictive analysis, data scientists' base on large sets of data to resolve with good veracity the outcome of social events like customer predilection, customer lifetime value, charismatic prices, and churn rates as it relates to businesses. This is, however, not limited to business insights as almost any event can be predicted with the right data analysis whether it is social opinions or investment markers.

And due to the demoralized nature of blockchain and the huge computational power available through it, data scientists even in smaller organizations can undertake comprehensive predictive analysis tasks. These data scientists can use the computational power of disparate thousand computers connected on a blockchain network as a cloud-based service to analyze social outcomes in a scale which would not have been otherwise possible.

5.2.4 REAL-TIME DATA ANALYSIS

As has been advertised in financial and payment systems, blockchain makes for real-time cross border transactions. Several banks and finch innovators are now scrutinizing blockchain because it affords fast—actually, real-time—settlement of huge sums irrespective of geographic barriers.

In the same manner, organizations that desire real-time analysis of data in large scale can call on a blockchain-enabled system to achieve. With blockchain, banks and other organizations can examine changes in data in real time making it possible to make quick decisions—whether it is to block a suspicious transaction or track abnormal activities [9].

5.2.5 Manage Data Sharing

In this regard, data received from data studies can be stored in a blockchain network. This way, project teams do not reiterate data analysis already carried out by other teams or unlawfully reuse data that's already been used. Also, a blockchain platform can help data scientists fabricate their work, probably by trading analysis outcomes stored on the platform.

5.3 HOW BLOCKCHAIN AND BIG DATA COMPLEMENT EACH OTHER?

Blockchain and big data are two technologies in full swing, but they are also two parallel technologies. In a decade, the blockchain is becoming the heart of computer technologies. Blockchain is a cryptographically secure distributed database technology for storing and transmitting information. Every record in the database is called a block and contains details such as the transaction date and a link to the previous block. The main advantage of the blockchain is that it is a decentralized system.

5.3.1 Data Quality and Privacy

In big data technology, a large amount of data is generated daily;analyzing and making use of this huge amount of information is the top priority for all kinds of businesses. However, the most important problem that hinders the unanimous adoption of big data is the lack of security and privacy protection.

The increase in the use of big data on the internet has led to the explosive growth of data size. However, the trust issue has become one of the biggest problems of big data, leading to the difficulty in safe data circulation and industry development. The blockchain technology provides a solution to this problem by combining non-tampering, traceable features with smart contracts that automatically execute default instructions [10].

Cryptocurrency such as Bitcoin, Ethereum, and the blockchain can support any type of digitized information; this is why it is possible to use it in the field of big data, especially to increase the security or the quality of the data.

For example, a hospital can use it to ensure that patient data is kept safe, up-to-date and that its quality is fully conserved. By storing health databases on the blockchain, hospitals ensure that all its employees will have access to a single, unchangeable source of data as shown in Figure 5.5.

Certainly, poor data management in the health care environment carries a risk that the patient may be mishandled, misdiagnosed, or the results of their tests may be lost or corrupted. Similarly, two physicians who are examining the same patient may have access to two different sets of data. The blockchain eliminates the risk [11].

5.3.2 Data Transparency and Automation

Blockchain is a new filing system for digital information, which stores data in an encrypted, distributed ledger format. This is because data is encrypted and distributed

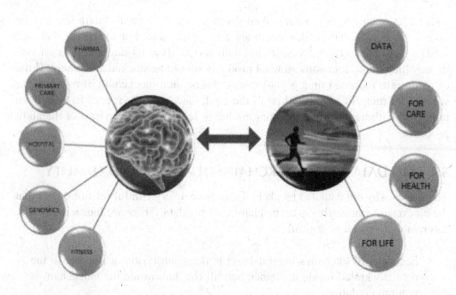

FIGURE 5.5 Hospital management system.

over many different computers, it enables the creation of tamper-proof, highly robust databases that can be read and updated only by those with permission [12].

The use of the blockchain also raises issues of confidentiality, in direct contradiction to why this technology originally became popular. Several experts are concerned that transaction records may be exploited to build consumer profiles or other misuses. However, blockchain enhances the data transparency, if an entry cannot be verified, it is automatically rejected. The data is therefore completely data transparent. Other experts are also concerned about the impact of blockchain and big data on the environment.

Blockchain along with big data technology helps in prevent possible data leaks. Once the information is stored in the channel, even the most senior managers of the company will require multiple permissions from other points in the network to access the data [13]. It is therefore nearly impossible for a cybercriminal to seize it. The blockchain also allows sharing data more serenely. By using the example of the hospital, an institution may need to share health data with the courts, with insurance companies, or with the employers of a patient. However, without the blockchain technology, this procedure can present risks.

Blockchain can help us track, understand and explain decisions made by artificial intelligence (AI). Artificial Intelligences based decisions can sometimes be hard for humans to understand because they are capable of assessing a large number of variables independently of each other and extracting the ones that are important to the overall task it is trying to achieve.

As an example, AI algorithms are expected to frequently be used in making decisions regarding whether financial transactions are fraudulent, and should be blocked or investigated. Even sometimes it is necessary to have these decisions audited for accuracy by humans. And given the huge amount of data that can be taken into consideration, this becomes a complex task. For example, Walmart feeds a month's worth of transactional data across all of its stores into its AI systems, which make decisions on what products should be stocked and where. If the decisions are recorded on a point-by-point basis, on a blockchain, it makes it far simpler for them to be audited, with the confidence that the record has not been tampered with between the information being recorded and the start of the audit process.

5.4 BIG DATA AND BLOCKCHAIN: QUANTITY AND QUALITY

The reason why big data and blockchain can have a very fruitful relationship is that the blockchain can easily cover the blemish of big data. There are four reasons why this relationship can be fruitful:

- Security: Blockchain's biggest boon is the security that it imparts to the data stockpiled inside it. Remember, all the data inside the blockchain is non-manipulative.
- Transparency: The transparent architecture of the blockchain can help you shred data back to its point of origin.
- Decentralization: All the data that is stored inside a blockchain is not owned by a single entity. So, there is no chance of loss of data if that entity gets compromised in any way.
- Flexibility: The blockchain can store all kinds and different types of data. If you consider all these factors, the conclusion can be drawn is that whatever data comes out of the blockchain is valuable and non-tamperable. It has already been cleaned and is fraud-proof. That is a potential bonanza that many companies are looking to exploit.

5.4.1 DECENTRALIZATION

Before Bitcoin and bit torrent came along, we used centralized services. You have a centralized entity which stores all the data and you'd have to interact merely with this entity to get the information you required.

Another example of a centralized system is banks. They stockpile all your money, and the only way to pay someone is by going through the bank. The traditional client-server model is an ideal example shown in Figure 5.6.

When you Google to search for something, you send a query to the server who send the relevant information getting back to you. That is simple client-server.

Since centralized systems have treated us well for many years, they have several vulnerabilities.

- First, as the system is centralized, all the data is stored in one spot. This makes in easy target spots for potential hackers.
- If the centralized system needs to go through a software upgrade, it would halt the entire system.

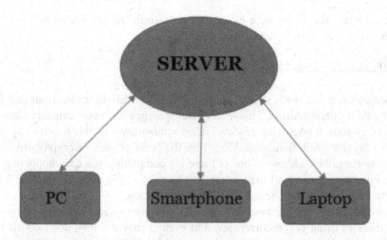

FIGURE 5.6 Client-server model.

- What if the centralized entity shut down for whatever reason? In this case, nobody will be able to access the information that it possesses.
- Worst case scenario, what if this entity gets corrupted and vicious? In this case, all the data that is inside the blockchain will be compromised.

So, what happens if we remove this centralized entity?

In a decentralized system, the information is not stored in one single entity. In fact, everyone in the network possesses the information.

In a decentralized network, if you want to interact with your friend then you can directly interact with them without going through a third party. That's the main ideology behind Bitcoin. You alone are in charge of your money. You can send your money to anyone you want without going to a bank (Figure 5.7).

5.4.2 TRANSPARENCY

Transparency is one of the most important factors in blockchain technology which is somehow misunderstood. Some people say that blockchain gives you privacy whereas some say that it is transparent. Why do you think that happens? Well, a person's identity is hidden via obscure cryptography and represented only by their

FIGURE 5.7 Example of data transaction.

public address. So, if you were to look up a person's transaction history, you will not see "Bob sent 10 BC" instead you will see:

IMF1bhsFLkBzzz7vpFYEmvwT2TbyCt7NZJ sent 10 BC.

The person's real identity is secure, but you will still see all the transactions that were done by their public address. This level of transparency has never existed before in a financial system. It adds the extra level of accountability which is required by some of these biggest institutions. Speaking from the point of view of cryptocurrency, if you know the public address of one of these big companies, you can simply pop it in an explorer and monitor all the transactions that they have engaged in. This forces those to be honest, as no one can deny their transaction.

However, that's not the best use case. We are pretty sure that most companies don't transact using cryptocurrencies, and even if they do, they don't do all their transactions using cryptocurrencies. What would happen if the blockchain technology was integrated, say in their supply chain? You can see why this technology can be very helpful for the finance industry?

5.4.3 IMMUTABILITY

Immutable, in general, means data cannot be changed once entered. So, immutability, in context of the blockchain, means that once details have been entered into the blockchain, it cannot be modified.

In simple terms, if we talk about hashing, it means taking an input string of any length and giving out an output of a fixed length. In cryptocurrencies like Bitcoin, the transactions or records are taken as an input which is then executed through a hashing algorithm (Bitcoin uses SHA-256), which gives an output of a fixed length which act as a signature.

5.5 BIG DATA CHALLENGES

Data volumes are persisting to extend and so are the probabilities to do with so much raw data available. However, organizations need to know what they can do with that data and how much they can resist building enlightenment for their consumers, products, and services. While big data offers a ton of advantages, it comes with its own set of concern and challenges. This is a new set of complex technologies, while still in the emergent stages of development and evolution.

Some of the commonly faced issues include incompetent knowledge regarding the technologies involved, data confidentiality, and inadequate analytical capabilities of organizations. Many enterprises face the issue of a lack of skills for dealing with big data technologies. In recent days, people aren't trained to work with big data, which may become a bigger problem.

This isn't the challenge or problem though. There are other challenges too, some are distinguished after organizations begin to move into the big data space, and some while they are cobblestone the roadmap for the same. Some of the challenges of the big data are discussed in the following sections.

5.5.1 HANDLING A LARGE AMOUNT OF DATA

Definitely one has to be concern about size when it comes to big data. Managing large data and rapidly increasing volumes of data has been an issue for decades. This challenge was earlier directed by developing faster processor to endure with increasing volume of data. However, data volume is scaling faster than computer resources and CPU speed is passive, instead of processors doubling their clock cycle frequency every 18–24 months. Due to power restraint, clock speed has largely stalled and processors are developed with more cores.

There is a huge detonation in the data accessible. Look back a few years and compare it with today, and you will see that there has been an exponential increase in the data that enterprises can access. They have statistics for entire, right from what a consumer likes, to how they react, to a particular track, to the amazing restaurant that opened up in Italy last weekend.

This data exceeds the amount of statistics that can be accumulated and computed, as well as reclaimed. The challenge is not so much the possibility, but the management of this data. With data claiming that statistics would increase 6.6 times the distance between earth and moon by 2020, this is definitely a challenge.

Along with rise in unstructured or disorganized data, there has also been a rise in the number of data formats. Video, audio, social media, smart device data, etc., are just a few to name.

Some of the contemporary ways developed to manage this data are a hybrid of relational databases connected with NoSQL databases. An example of this is MongoDB, which is an innate part of the MEAN stack. There are also dispersed computing systems like Hadoop to help manage big data volumes.

Netflix is a content streaming platform based on Node.js which allows us to view different series with the payment of its charge. With the enlarged load of content and the complex formats available on the platform, they need a stack that can handle the storage and recuperation of the data. Hence, they used the MEAN stack, and with a relational database model, they can manage the data.

5.5.2 SECURITY ASPECTS AND CONSTRAINTS

A lot of organizations marked allegation that they face trouble with data security. This happens to be the next bigger challenge for them than many other data-related problems. The data that comes into venture is made available from a wide range of authority, some of which cannot be trusted to be secure and docile within organizational standards.

They need to use a variety of data collection approaches to keep up with data needs. This in turn leads to inconsistencies and inaccuracy in the data, and then the outcomes of the analysis. An example such as annual turnover for the retail industry can be different if analyzed from different onset of input. A business will need to accustom the differences, and narrow it down to an answer that is valid and interesting.

This data is made available from various sources, and therefore has potential security problems. You never know which channel of data is compromised, thus compromising the security of the data available in the organization gives a chance to hackers to move in. Henceforth, it's necessary to introduce data security best practices for secure data collection, storage and retrieval and eliminate inconsistency of data.

5.5.3 Efficiently Processing Unstructured and Semi-Structured Data

Databases and warehouses are disappointing for processing of unstructured and semi-structured data. With big data, read or write operations are highly concurrent for large number of users. As the size of database enlarges, algorithm may become insufficient and invalid. The CAP (Consistency, Availability, Partition tolerance) theorem states that it is impossible for a dispersed system to have all the three; we can choose only 2 out of 3.

The fact that is difficult to serve such kind of data in a rigid form makes it difficult to process, which leads to the announcement of new processing mechanisms such as NOSQL. It is worth noting that the definition of big data is continuing changing to include new minutiae, which are becoming very important to consider. A good big data is measured by the following 5Vs:

1. Variety
2. Volume
3. Veracity
4. Value
5. Velocity

Data must be structured as a first step in data analysis. Example: A patient in a hospital who has one record for medical report/lab test, one for surgical operations, one for each admission at the hospital, and one for a lifetime hospital interaction with the patient. The number of surgical operations and lab tests per record would be different for every patient. The three design choices listed have consecutively less structure and conversely, consecutively greater variety. Pointing out the acquisition of big data from numerous sources with variety of structures, structuring these data is almost impossible before data analysis. Thus, it is one of the next important challenges of big data.

5.5.4 Computations

The main idea behind big data is to extract useful acumen by performing specific computations. However, it is important to secure and protect these computations to avoid risk or attempt to change or skew the extracted results and avoid loss of data. It is also important to protect the systems from any endeavor to spy on the nature or the number of performed computations.

5.5.5 Communication

Big data is hoarded in several nodes belonging to many clusters which are scattered all over the world. All communications between clusters and nodes are ensured through ordinary public and private networks. However, if someone can customize the inter-node communication it would be easy to extract valuable information. Therefore, it is a good challenge for big data tools to adopt new secure network protocols in order to protect synergy between different parties.

5.5.6 ACCESS CONTROL

In a big data context, approach to the data should be managed by a strong access control system to revoke any malicious party from getting access to the storage servers. That is, the only node with sufficient administrative rights could have the probability to manage and process any content. Furthermore, any modification in clusters' state such as addition or deletion of nodes should be examined by an authentication mechanism to protect the system from malicious nodes.

5.5.7 RANDOM DISTRIBUTION

The concept of big data analytics is mainly based on parallelism, for this, the large data is gathered and processed in different clusters, which are a set of dispersed servers around the world and acting as one powerful station. The main issue with this topology is it is very hard to know the exact location of storage and processing which can conclude in many security problems and regulation breaches. The main challenge with big data solutions is to be able to disperse storage and processing according to the regulations and data sensibility.

5.6 SOLUTIONS TO BIG DATA USING BLOCKCHAIN

The popularity of blockchain technology and its application results in much ongoing research in different practical and scientific areas. Although still the new and the in experimenting phase, the blockchain is being seen as a revolutionary solution, addressing modern technology concerns like decentralization, trust, identity, data ownership, and data-driven decisions. While at the same time, the world is facing an expansion in the quantity and diversity of digital data that are generated by both users and machines. In actively searching for the best way to store, organize and process big data, the blockchain technology comes in providing significant input. The solution to decentralized management of private data, digital property solutions, Internet of Things-based communication and public institution's reforms are having a numerous impact on how big data is evolved [14].

5.6.1 IMPLEMENTATION OF SHA 256 WITH BLOCKCHAIN FOR BIG DATA TECHNOLOGY

Hashing is an important technique in blockchain technology. It is a mathematical process that is used for writing new transactions into a blockchain. The process is executed by a hash function with the help of a hashing algorithm (SHA 256). A hash function is a function in which input of any variable length of data or string will give an output of a fixed length. The output that we get from a hash function is known as hash. In hash function, the size of the input is not a matter; whether it is 2 or 2000, it will give the output of the same length. For example, if we use the SHA 256 algorithm for hashing it will always produce an output of 256-bits length.

For a hash function to be secure, it must have the following features:

- Deterministic: Which means the output or the hash should be the same even if the user execute the same input two or more times.
- The hash function using should be able to produce the output or hash as quickly as possible.
- For every hash—say H(f)—it should be infeasible to find an input f from H(f). Suppose we are using a 128-bit hash where the data is very huge and the Brute Force method is the only way to find the original input. In the Brute Force method, a random input is selected and hashed and is compared to the target hash. This process is repeated until it matches the input. Generally, it is not a practical method as there is a huge volume of data.
- For every small change in the input, it should make a huge change in the hash.
- The hash values should be unique, for example, for any two inputs A and B, the output hashes of H(A) H(B) should not be equal.

For every output Y, it should be infeasible to find an input X, such that

$$H\left(k|X\right) = Y$$

where k is a random value of high minimum entropy.

$$\left(k|X \text{ is the concatinaton value of } X \text{ and } k\right)$$

There is a number of hash functions available in the blockchain technology most commonly used once are:

SHA 256: Used in Bitcoin

Keccak 256: Used in Ethereum

5.6.1.1 Data Structure

Basically, data is stored in blockchains then, of course, it will have a data structure also. There are mainly two data structures used in blockchain one is pointers and other is linked list. Pointers are variables that store the address of the other variables as illustrated in Figure 5.8. In blockchain technology, more than the address the pointers also store the hash value of the previous block.

In the data structure level view, we can say that blockchain is basically a linked list in which each node stores a hash pointer and a data header. While the data header

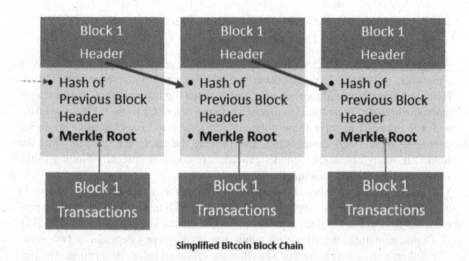

Simplified Bitcoin Block Chain

FIGURE 5.8 Simplified blockchain structure.

will store the data of that block, the hash pointer will have the address of the preceding block as well as the hash value.

5.6.1.2 Mining Process

Mining is simply a process in which new blocks are added to the blockchain. The miners verify a transaction that has been pushed to blockchain and add it to the blockchain if it is a valid transaction.

Each blockchain network has a time limit for the creation of a block (In Bitcoin, it is 9–10 minutes as of now). If the blocks are created at a faster rate, it will result in the generation of more hash functions in a short time which may result in the collision of hashes. Also, when new blocks are created faster, certain blocks shall not be the part of the main chain as more blocks are added simultaneously to the chain. So, to avoid this problem certain difficulty level is set for the hash. If we take the example of Bitcoin, when a new block arrives, the hash function hashes all the contents of the block. Later, the hashed output is concatenated with a nonce (a random string). Now, the entire concatenated string is hashed again and performed a difficulty level comparison.

If the new output is less than the difficulty level, the block is added to the chain; otherwise, the nonce is changed and the process is repeated until it passes the standards.

In Bitcoin, for every output W, it is infeasible to find an input X such that:

$$H(k|X) = W$$

where:
k = nonce,
X = hash of previous block, and
W = difficult level.

5.6.2   IMPLEMENTATION OF DIGITAL SIGNATURE-BASED SECURITY IN BIG DATA MANAGEMENT

A digital signature is a technique used to validate the authentication and integrity of a message, software, and digital document by using certain algorithms like Diffie–Hellman key exchange algorithm, etc. Similar to the digital equivalent of a handwritten signature or stamped seal, a digital signature offers far more inherent security, and it is intended to solve the problem of tampering and impersonation in digital communications. Digital signatures also provide the assurance, by validating the sender. In many countries, digital signatures are considered legally binding in the same way as traditional document signatures.

Digital signatures are based on public-key cryptography technique, also known as asymmetric cryptography. Applying a public key algorithm, such as RSA, one can generate two keys that are mathematically linked: one private key and one public key [15].

Digital signatures work because public key cryptography depends on two mutually authenticating cryptographic keys that are validated before decrypting. The person who is creating the digital signature uses their own private key to encrypt signature-related data; the only way to decrypt that data is with the sender's public key. This is how digital signatures are authenticated.

5.6.3   PUBLIC KEY INFRASTRUCTURE (PKI) FOR BIG DATA SECURITY

A PKI is a system in which we can trust the third-party user to identity inspection and assurance which is done basically by a certificate authority with the use of cryptography involving private and public keys. A typical PKI system consists of the following:

- Client software system
- Certificate authority server
- May involve a few smart cards
- Operational methods

Public key cryptography is an advance field in IT security that enables the trade-off between entities that are confidential in an open network. Public key cryptography enables the protection techniques that don't exist in the traditional cryptography, importantly in the digital signature.

The public key cryptography consists of a public and a private key but does not require a confidential exchange of secret keys. Public key cryptography still is of vital importance as the public key is verifiably authentic, and the private key remains private. This is known as public key infrastructure (PKI), which manages the key pairs [16,17].

5.6.3.1   Authentication of Public Key Infrastructure

Firstly, the certificate authority will check the user, different certificate authorities have different identity validation process. The certificate authority acts as a trustee as well as is an independent provider of digital certificate. In this scenario, some of them may grant access to users because of a digital certificate, which will contain only the name and email address. While the certificate authority users can include personal interviews, background checks, etc. If the user is granted a digital certificate, then they will have two components, i.e., private key and public key.

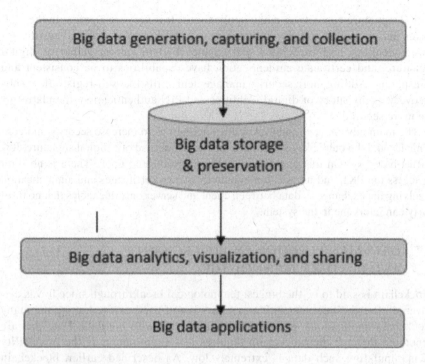

FIGURE 5.9 PKI model.

For example, the user wishes to send an email to his business associates and wants to sign digitally the email with his private key. After signing the email, the user sends it to his business associates. Then the business associate will decrypt the email from the user's public key. This example of digital certificate provides secret information that can be shared with user authentication, without exchange the secret key in advance. The PKI is also used in medical application systems (Figure 5.9).

5.6.3.2 Uses of PKI

Public Key Infrastructure (PKI) is a system designed to manage the creation, distribution, identification, and revocation of public keys. The uses of PKI are listed below.

Encryption or sender's authentication of an email message and other documents

- Authentication of users to applications like for SMART CARD login system
- Bootstrapping that enables secured communication with some protocols, such as internet key exchange

Integrated security infrastructure for encryption, digital signature, and Certificate authority:

A PKI maintains a trustworthy networking environment, established by providing keys and certificate management services that enable encryption and digital signature capacity in many applications. It is very transparent in the platforms and is easy to use. By managing the full life cycle of digital certificate-based identities and encryption that enables PKI to entrust authority, digital signature, and certificate authentication have capabilities to be consistent and transparent. Adding more security management skills like self-registration, self-recovery or inventory of digital identities and PIN authentication, the platforms are more secured.

The main advantage of public-key cryptography is to increase security and convenience and the public key system is to provide a method of a digital signature. It is a trust-based system that enables its customers to depend upon. There is no limit to access the PKI, and users can maintain their own certificates and authentication involving the exchange of data between client and server only. It enables that no third party can intervene in the system.

5.7   DATA MANIPULATION OF BLOCKCHAIN AND BIG DATA TECHNOLOGY

Blockchain is said to be the biggest technological breakthrough since it was discovered after the internet, with an endless list of used cases. Among the many blockchain applications, one of special interest is with big data. Data that are generated through the blockchain requires less examination as the probability of manipulating such data is extremely low. As described earlier, blockchain combines hashing and cryptography techniques to create data stored in a series of blocks which is almost immutable due to the tremendous amount of computing power that would be required to manipulate the original state of the block. Centralized data is insecure with several reports of data hacks; it is obvious that data in centralized storage are at the mercy of hackers and other malicious entities. The good thing is that blockchain aims to restore power into the hands of the user; there are unlimited opportunities that exist for big data with the application of blockchain. An ideal world is one where data can be managed by the individuals generating the data. Big data generated through blockchain technology is growing over regular big data and will eventually be the new way of generating data [18].

5.7.1   MEDICAL RECORD SYSTEM

Applications of blockchain are now extended in the field of healthcare information technology. The medical record includes text reports, word documents, images, videos, and other data. They all are important in medical history. Central storage is important as many doctors may not have their medical record systems. The main problem with medical records is the increasing storage due to the forever increasing number of reports or files is caused by the continuous contribution of reports from doctors and patients. Medical record ownership and record sharing among doctors is extremely controversy within the medical community. Many doctors worry about possible lawsuit initiation by others or even by the patient if things go wrong. Medical-related legal matters or lawsuit may incur at any time. Therefore, all documents generated must be properly stored and must be produced. Data security is important, as it is necessary to protect the privacy of both patients and doctors.

Blockchain can prevent duplication of transactions by accepting the very first arrived transaction for a specific record and rejecting other transactions. Duplication of data can be avoided and can be detected immediately through the digital signature of the duplicated records. The blockchain mechanism can be implemented directly on the medical reports, database, backup, and duplicated hardware that have been used to protect data from the risk of tampering. Hash values and the digital signature of each record can further strengthen protection from the tampering of data.

Blockchain mechanisms can be implemented for all the encrypted medical records and files, encryption of medical records and files is needed for further protection from unauthorized access. The digital signature can be used for integrity checks of the original medical record or files so that each record or file is not being deleted or tampered, each encrypted record or file should be included serially inside the blockchain implemented.

Blockchain is a good choice for present and previous records management system which concentrates on the integrity of medical reports. However, in the long run, blockchain-related application development may hit a solid wall very soon when several medical data records or files hit a larger volume. Blockchain can protect the integrity; however, it doesn't guarantee the existence or the interpretation of the reports, it cannot control the viewing right, it cannot solve the challenges of outdating formats, it cannot handle forever increasing storage size, and it cannot handle deletion or purge [19].

5.7.2 BANKING AND MANAGEMENT

Blockchain technology is interrupting the banking industry and contributing to the increased big data in banking. There exists a gap in research and development of blockchain with big data technology in banking sector, and this gap is expected to have a significant negative impact on the adoption and development of blockchain technology for the banking sector [20], as shown in Figure 5.10.

The existing relationship between the banking industry and blockchain is complex because beyond the numerous opportunities for streamlining traditional banking processes, blockchain is viewed as a threat to established models [21].

The verification of the authenticity of the consumer's identity is a frequent and important task for the banks owing to anti-money laundering regulations. The ever-increasing problems of terrorism, Know Your Customer (KYC) is a crucial aspect related to the prevention of the criminal use of banking funds and services in the form of money laundering and terrorism.

Security is of supreme importance to the banking industry. Blockchain can offer magnified security for banking data as historical information cannot be altered and the new information which is added in real-time is essentially shared by multiple entities and thus making it difficult to manipulate the data. Any alterations to data within a block can be tracked and monitored to prevent fraud and misuse, and more importantly blockchain technology allows for real-time communication and updating on potential scams.

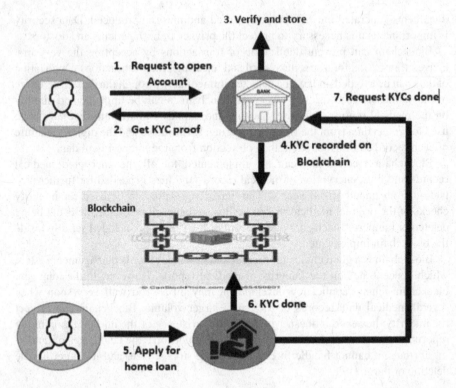

FIGURE 5.10 Blockchain use in the banking sector.

Blockchain is also affecting the antecedents of trust including confidence, integrity, reliability, responsibility, and predictability. The transparency in blockchain's distributed ledger is also useful as it would enable regulators to easily scrutinize financial actions. As a trusted network, consumer data stored via blockchain can only be accessed by trusted sources in addition to the data security attainable via the encryption capabilities of blockchain. Blockchain can also achieve simultaneous security and privacy by enabling confidentiality through public key infrastructure and by maintaining the size of a ledger.

The security element features as both an opportunity and a challenge for the adoption of blockchain in banking. It is no secret that banking requires high security. Historically, banks had to only be concerned with protecting their safes and deposits in high street banks. The move to online banking saws the emergence of a new security threat from hackers around the globe, thus creating a job market for cybersecurity and increased costs for banks.

Data privacy remains a concern within blockchain technology as transactions can be seen throughout network nodes that produce metadata leading to pattern recognition. Another security challenge posed by blockchain is anonymity. As blockchain technology allows for anonymity, such untraceable transactions would threaten the banking sector and the regulator as it increases the difficulty of tax and can also aid criminals with money laundering.

5.7.3 SOCIAL MEDIA DATA

The description of blockchain is described differently by everyone some describe blockchain as an improvised version of the internet. The blockchain technology is designed to promote user responsibility and data security that is lacking in how data is handled on the internet at present [22]. The primary purpose of blockchain was only for Bitcoin and its use in finance, it soon changed social media platforms and how online business is conducted. Blockchain-based social media will improve the transparency and security of data. Blockchain's immutability will make identity verification of users easier and prevent identity theft as users without proper authorization's access will be denied automatically. It can also reduce fake content, spambots, and fraudulent ad impressions, etc.

Blockchain has enforced social media platforms to reorganize their sharing algorithm, data security policy, and payment methods to increase the security of data. For large firms and leading companies that rely on social advertising, blockchain technology has entailed fast thinking on how to adopt crypto-transactions in their business.

Nowadays everyone is having more than one account on social media to interact with friends and family. The social media platform has full access to the personal data and information that is provided to the social media platforms. The social media platforms can give information to stakeholders or to anyone who pays a huge sum of money. They usually need the information to create highly targeted ads and campaigns. At this point, the safest way to take full control is to make a switch to blockchain-based social media platforms, which are the ultimate solution to privacy and data security problems. These decentralized versions of social media channels have strong end-to-end encryption; users can rest easy knowing that no digital breach will be left behind for the third party to pick up. These blockchain-based social media platforms own digital currencies that can be used to process in-platform transactions to make the system more secure.

5.7.4 SPACE DATA

The existing space tech and space development sector is still dominated by national and supranational space programs like NASA, the Canadian Space Agency, the European Space Agency (ESA), China's CNSA, and Russia's Roscosmos. They have their means of secure, reliable communications and will not share crypto enthusiast's desire for eluding national authority.

These agencies are noticeably cautious. A recent study by the ESA on earth observation (EO) pointed out that distributed ledgers are not yet considered to be immediately applicable to space systems operations and EO data exploitation. They are clear that they believe that there is a need to demonstrate how blockchain could help solve existing challenges for the EO sector, that many of the benefits of blockchain are theoretical, and that scalability is a large concern.

The EO is growing due to the benefits of big data applications, artificial intelligence (AI), and machine learning. Machine-to-machine (M2M) communications within networks are becoming more routine, and data or information network flows are becoming less linear and more dynamic.

The ESA believes that the growth of these automated dynamic networks may create a need for a secure, definitive reference for data distribution and tracking, as well as the record of processing steps. The traceability and immutability of these kinds of transaction records will be precious, as will those of mission planning and operations, digital supply chains, and physical supply chains in EO manufacturing. This is especially important as EO and other space operations become more vulnerable to cyberattack.

The ESA has also been involved in events focused on supporting space-based blockchain applications and recognizes that its role may well be in creating a bottom-up environment for private-sector innovators.

The NASA engineers have expressed similar views on how blockchain could provide interference-resistant and reliable networking. With the FAA's requirement that all aircraft flying in the American National Airspace System adopt the automatic dependent surveillance system (ADS-B), both military and civilian aviators are becoming increasingly concerned about the security issues with ADS-B's lack of a robust security model. Both are concerned about third-party spoofing and about its unencrypted plaintext broadcasts.

NASA proposed a blockchain-based solution that the complex hyper ledger fabric solution differs from fin-tech style, coin-based blockchain networks. It still provides the immutability and transparency needed, however, while also respecting the privacy and security needs of military aviators [23].

5.8　SCENARIO BASED ANALYSIS OF BLOCKCHAIN TOWARD BIG DATA

What makes blockchain an irresistible offering for data scientists is the power it holds against more traditional databases. The perfect database would be unalterable, histories by nature, and won't permit anyone or anything corrupt or modify its registers. Blockchain makes this possible. There are three ways in which blockchain is disrupting big data analytics, as discussed in the following sections.

5.8.1　ACCESSIBILITY TO DECENTRALIZED TOOLS

One of the main barriers integrating big data analytics into existing infrastructure is the high cost associated with it. In recent, big data was something only large corporations could use and leverage for their performance. But with the advent of subscription-based cloud analytics software and business intelligent tools, this has changed for the better. But today, blockchain-based tools aim to expand this cost-efficient accessibility to data analytics tools by decentralizing the technology required.

Endor is a start-up working with multinational enterprises to expand their predictive analytics offerings. Recently, they announced a blockchain protocol, which is being regarded by many as the Google of predictive analytics. It clubs AI and analytics in a seamless manner so that even normal users can ask simple questions to get accurate predictions. The company further plans to bring together data providers, developers, and users to nurture a predictive analytics ecosystem built around blockchain technology. By relying on such decentralized tools, the cost of using predictive and big data analytics is going to go down.

5.8.2 NEW FORMS OF DATA MONETIZATION

Data is the single most-valuable commodity in the modern world and combining blockchain and big data can democratize the way data analytics is shared and monetized by completely removing the middle person. This, in turn, would lead to the consumers gaining stronger negotiation powers over businesses, while allowing them to control which business has access to their data and which does not.

Blockchain can also give an increase to new forms of data monetization because of the following changes it brings to the table:

- All businesses involved in a transaction would have access to the same quantity and quality of data, which would accelerate data acquisition, sharing, the quality of data, and data analytics.
- All transactions will be registered and kept on a single file, providing a complete overview of the transactions while eliminating the requirement for multiple powerful computer systems.
- Individuals will be able to control and manage their data without depending on a third party, therefore opening up the market.

Platforms such as Wibson are already encouraging data owners to share their data information with data consumers while getting paid to do so. Wibson's marketplace allows users to leverage their infrastructure to monetize their anonymous private information while being supported through a token-based economy. Because of the transparency afforded by blockchain, sellers can see how their information is being used even after the transaction has taken place.

5.8.3 FLUENT DATA EXCHANGE

Most large organizations leverage the amount of data they have to sell this data to others. In the case where your business has the latest and best analytics tool in the market, it will not be worth anything unless you have the data to feed into it. This costly access denies small businesses and research groups to work with large volumes of data, while also greatly stifling the way this data is exchanged.

To solve such issues, data exchange platforms such as Dock are enabling working professionals to manage their job profiles under a single platform. Instead of working and sifting through multiple profiles on multiple job sites, in this case, head-hunters can hunt through a single repository of validated, timely, and up to date information. Dock also helps consolidate certifications and other experiences gained from different platforms while storing all this data on the blockchain, allowing professionals to create in-depth profiles. According to Forrester, the research group, close to 73% of enterprise data goes unused for data analytics. But blockchain can help bring down these boundaries by making data exchange more secure and easy, without any major infrastructural costs associated with it.

5.9 APPLICATION OF BIG DATA WITH BLOCKCHAIN

There are many applications of big data with blockchain discovered in the today's world; some of them are presented in the following sections.

5.9.1 STROJ

Blockchain is a decentralized ledger of transactions that is accessed on a peer-to-peer basis. Every user in the network validates these transactions, so the ledger is secured and maintains integrity indefinitely. While it's commonly applied to cryptocurrency transactions and now smart contracts, virtually any data can be securely stored within the blockchain.

Enter decentralized data storage providers like Storj that could provide savings for big data, who are currently forking-out for traditional cloud storage. According to VentureBeat, a preliminary study revealed that the decentralized approach could reduce the costs of storing data by 90% compared to the likes of Amazon Web Services' Cloud solution.

The main benefits are privacy and security. Data is not open to a single point of attack, and the customer is not at the whim of the provider and its data centers. If the power goes down or a data center is corrupted, the algorithms ensure data is distributed widely enough to maintain high availability. Storj currently stores over 100 petabytes of data. If you need funds for data storage, you can always use a loan-matching service.

5.9.2 FILECOIN

Similar to Storj, FileCoin has the lofty goal of completely transforming how data is stored and securing and decentralizing the internet. Using the Egyptian government crackdown of the web as an example of existing points of failure, the developers envision a truly open blockchain-driven internet in the future.

Right now, that means offering data storage solutions and the ability to earn its corresponding FileCoins by using your spare space to mine. The market bidding process ensures competitive pricing.

5.9.3 OMNILYTICS

Omnilytics is a new technology that aims to combine the blockchain with big data analytics. It uses artificial intelligence and machine learning as part of this process, with marketing, financial due diligence, auditing, trend forecasting, and many other applications across industries. They believe that by utilizing blockchain technology they can disrupt the big data giants. states, "It powers our smart contracts, distributed data fingerprinting, data exchange and other protocols and APIs." Data partners can track the performance of their data and pricing remains competitive based on the usage of said data. The inherent openness and honesty of the network offer a new level of trust and transparency.

5.9.4 DATUM

Datum is a decentralized storage network driven by the data access token (DAT). It puts the focus on the individual, who can monetize their data in an open and honest marketplace, instead of being exploited by the current data giants like Facebook.

Instead of making corporation's money via your tacit agreement to use their services, you have full control and make money yourself. The blockchain also ensures that there are no breaches.

This doesn't shut the big players out and may give them great access, but it will be a fairer and more secure system.

5.9.5 RUBLIX

Rublix aims to unite cryptocurrency investors across the world with a simpler trading platform that verifies the authenticity and credibility of traders and provides access to market information to reduce the current confusion.

The majority of new investors are limited by the desktop and mobile tools they choose and are a victim of a wild-west approach. Rublix aims for an integrated platform that makes use of blockchain secured and verified investment data analytics, with the best tools, services, and information rising to the top.

5.9.6 PROVENANCE

Provenance aims to use the blockchain to build trust in the journey of a product. Customers get to know verified information about what the product is made of, where it came from, and its impact on the environment—producers and retailers benefit from better product-tracking and by empowering their customers with this new information.

Over time as the data builds, producers and retailers also get insight into exactly what customers want and can tailor their goods and services accordingly. Provenance's core is building transparency throughout the supply chain.

REFERENCES

1. "Introduction to blockchain technology: Set 2," Geeks for Geeks, August 14, 2018. [Online]. Available: www.geeksforgeeks.org/introduction-to-blockchain/.
2. D. Puthal, N. Malik, S. P. Mohanty, E. Kougianos, and C. Yang, "The blockchain as a decentralized security framework [Future Directions]," *IEEE Consumer Electronics Magazine*, vol. 7, no. 2, pp. 18–21, 2018.
3. A. A. Siyal, A. Z. Junejo, M. Zawish, K. Ahmed, A. Khalil, and G. Soursou, "Applications of blockchain technology in medicine and healthcare: Challenges and future perspectives," *Cryptography*, vol. 3, no. 1, p. 3, 2019.

4. "What is Blockchain? "Explained simple: Lisk academy," Lisk. [Online]. Available: https://lisk.io/academy/blockchain-basics/what-is-blockchain.

5. Q. Xu, K. M. M. Aung, Y. Zhu, and K. L. Yong, "A blockchain-based storage system for data analytics in the internet of things," *New Advances in the Internet of Things Studies in Computational Intelligence*, vol. 715, pp. 119–138, 2017.

6. Z. Zheng, S. Xie, H. Dai, X. Chen, and H. Wang, "An overview of blockchain technology: Architecture, consensus, and future trends," *2017 IEEE International Congress on Big Data (BigData Congress)*, 2017.

7. S. Arora, S. Sinha, M. Reaney, and J. McAllister, "Introduction to blockchain technology and what it means to big data," *Big Data Made Simple*, 2018. [Online]. Available: https://bigdata-madesimple.com/introduction-to-blockchain-technology-and-what-it-means-to-big-data/.

8. S. Sagiroglu and D. Sinanc, "Big data: A review," *2013 International Conference on Collaboration Technologies and Systems (CTS)*, 2013.

9. S. Sarikaya, "How blockchain will disrupt data science: 5 blockchain use cases in big data," Medium, January 10, 2019. [Online]. Available: https://towardsdatascience.com/how-blockchain-will-disrupt-data-science-5-blockchain-use-cases-in-big-data-e2e254e3e0ab.

10. L. Yue, H. Junqin, Q. Shengzhi, and W. Ruijin, "Big data model of security sharing based on blockchain," *2017 3rd International Conference on Big Data Computing and Communications (BIGCOM)*, 2017.

11. A. Sharma, "How Blockchain and Big Data Complement Each Other," [Online]. Available: https://hackernoon.com/how-blockchain-and-big-data-complement-each-other-92a1b9f8b38d.

12. J. Hwang, M.-I.Choi, T. Lee, S. Jeon, S. Kim, S. Park, and S. Park, "Energy prosumer business model using blockchain system to ensure transparency and safety," *Energy Procedia*, vol. 141, pp. 194–198, 2017.

13. A. Outchakoucht, H. Es-Samaali, and J. Philippe, "Dynamic access control policy based on blockchain and machine learning for the internet of things," *International Journal of Advanced Computer Science and Applications*, vol. 8, no. 7, pp. 417–419, 2017.

14. E. Karafiloski and A. Mishev, "Blockchain solutions for big data challenges: A literature review," *IEEE EUROCON 2017—17th International Conference on Smart Technologies*, 2017.

15. D. Salomon, *Data Privacy and Security*. New York: Springer, 2003.

16. M. Koushikaa, S. Habipriya, S.S. Aravinth, T. Karthikeyan, and V. Kumar, "A public key cryptography security system for big data," *IJIRST—International Journal for Innovative Research in Science & Technology*, vol. 1, no. 6, pp. 311–313, 2014.

17. X. Zhang, K. Chang, H. Xiong, Y. Wen, G. Shi, and G. Wang, "Towards name-based trust and security for content-centric network," *2011 19th IEEE International Conference on Network Protocols*, 2011.

18. C. Udensi, "Blockchain impact on big data," Medium, January 14, 2018. [Online]. Available: https://towardsdatascience.com/blockchain-impact-on-big-data-39b38da7f4a5.

19. P. T. S. Liu, "Medical record system using blockchain, big data and tokenization," in *Lecture Notes in Computer Science* (including subseries Lecture Notes in Artificial Intelligence and Lecture Notes in Bioinformatics), vol. 9977 LNCS, pp. 254–261, 2016.

20. H. Hassani, X. Huang, and E. Silva, "Banking with blockchain-ed big data," *Journal of Management Analytics*, vol. 5, no. 4, pp. 256–275, 2018.

21. G. W. Peters and E. Panayi, "Understanding modern banking ledgers through block-chain technologies: Future of transaction processing and smart contracts on the internet of money," *SSRN Electronic Journal*, vol. 14, no. 4, pp. 352–371, 2015.
22. A. Chakravorty and C. Rong, "Ushare user controlled social media based on block-chain," *Proceedings of the 11th International Conference on Ubiquitous Information Management and Communication—IMCOM 17*, Beppu Japan, January, 2017.
23. "Blockchain and space: Chains in space," SpaceQ, 2019. [Online]. Available: http://spaceq.ca/blockchain-and-space-chains-in-space/. [Accessed: August 19, 2019].

6 Blockchain Storage Platforms for Big Data

Arnab Kumar Show, Abhishek Kumar, Achintya Singhal, S. Rakesh Kumar, and N. Gayathri

CONTENTS

6.1 INTRODUCTION

Blockchain is not a new terminology. It has been in existence for the last two decades; therefore, lots of new things innovated using blockchain. This chapter is about the use of blockchain technology to store big data, which is the new terminology and is the compilation of massive datasets. Since the data is extensive and complex, it can't be processed by the traditional data processing methods. These data sets can be used for an estimate the behavioral patterns. The demand for big data analytics is very high [15].

A huge amount of data can be used with blockchain technology at less time. Blockchain has the limit concerning putting away large measures of information, and that took over significant lots of time. Organizations can settle on putting away the information on a decentralized system structure with the utilization of blockchain [7]. Inferable from these advancements, endeavors won't need to bring about expenses for information of large stock of data on that stage. An extra advantage of utilizing the blockchain to deal with big data is the utilization of smart contracts usefulness. By coding the vital data, intelligent contracts can perform exchanges consequently. This can majorly affect lessening exchange costs.

6.2 SWARM-MUTABLE RESOURCE UPDATES-ENS-PSS

Swarm is a distributed database that stores and share data.The main objective of this storage platform is to provide decentralized and same information of end-users. It is a peer-to-peer network communication between nodes [8,9]. The report of this storage platform [1] can be fetched by anywhere. Data availability is free of this storage platform. There is no need of any server for accessing those data.

The main aim of Swarm's storage platform is to provide independent data accessibility for all end users. From the end user's view, Swarm is the same as the Web, with the exception that there is no need for hosting another server. There are two main differences between Swarm and other decentralized distributed storage platforms [2]: Where other platforms like BitTorrent, Zero net, and IPFS have to register and share the content, you host on the server, but a decentralized cloud storage service has been provided by Swarm storage platform.

Swarm is a peer-to-peer communication of network links that provide distributed services of resources such as storage, message forwarding, and payment processing with the peers. Everyone can easily connect to the network of Swarm client node on their server, desktop, laptop, or mobile device. The Swarm uses SWAP (Swarm accounting protocol) to perform the smooth and speedy operation. The Swarm customer is a part of the Etheriumstack. The implementation of Swarm reference execution is written in go-language and found in go-Etherium store.

Swarm is a collection of hubs or nodes of the developer P2P network. All nodes are running in the same network id and follow buzz protocol. For domain name, resolution Swarm nodes are connected to one Ethereum blockchain and for bandwidth and storage compensation it uses another Ethereum blockchain. Swarm use the same blockchain for transactions. A swarm system is distinguished by its system id which is an integer.

6.2.1 PUBLIC GATEWAYS

Swarm offers an HTTP intermediary programming interface that can use to cooperate with Swarm. The Ethereum establishment is providing an open gateway, which permits free access so individuals can easily access to Swarm network without running their node. This public gateway can be found on the Swarm website.

6.2.2 UPLOADING AND DOWNLOADING DATA

Swarm network is a service that provides APIs to upload and download content. This content is uploaded to Swarm cloud through URL. The Swarm service provider is using a decentralized peer-to-peer distributed infrastructure for uploading ad downloading materials. For smooth delivery, Swarm network is using SWAP protocol.

6.2.3 ETHEREUM NAME SERVICE (ENS)

The ENS offers a protected and decentralized way to address resources both on and off the blockchain. ENS is based on smart contracts on the Ethereum blockchain, meaning it does not suffers from the experiences of the DNS framework. ENS works in a distributed fashion for its frame and administration.

6.2.4 POSTAL SERVICE OVER SWARM (PSS)

The PSS is a messaging protocol that uses on the top of the Swarm network. The PSS can be stated as postal service over Swarm. This is a messaging protocol in the new system of internodes communication services. The PSS combines of the routing protocol and the encrypted messaging protocol.

6.2.5 SWARM MUTABILITY

Variable asset updates is a profoundly trial highlight, accessible from Swarm Point of Care Communication Council (POC3). It is under dynamic improvement, so anticipate change. If we want to change the information in swarm platform, the digital signature (hash value) will be changed in unusual ways. All the end users always want a modified version of a website or application with extra features. For this reason, Swarm accommodate with the Etherium Name Service (ENS) on the Etherium blockchain for easy to accessing the changing content.

6.3   INTER PLANETARY FILE SYSTEM

The internet-planetary file system (IPFS) is a distributed file system. This protocol uses a peer-to-peer network communication for storing and sharing information in a distributed file system [10]. The IPFS suggests their clients to get as well as to host content. As hostile a centrally arranged server, it is built around a decentralized network of client operators. Any client in the system can serve a document by its content address, and different peer in the network can discover and demand that content from any hub that makes them utilize a distributed hash table. To serve information on the Web, this protocol is used. At the present scene of the world, the Web works off location-based addressing where we go to a URL like something.com which has an IP address of A.B.C.D, and then we get served our articles. These URLs are indicated to specific servers around the globe. Instead, what IPFS does is it helps data dependent on what it is as opposed to where it is (location). With their routing algorithms, we can pick where we get our content from, and we can set our privacy and security of what peers/hubs we trust to get our records which is fascinating [11].

- By using hash addressing, the content will be immutable. Doesn't vanish like presently HTTP convention.
- It collects the contents from multiple hubs, instead of from one server, to save bandwidth.
- Access to the offline content or in low availability third-world or rural areas.
- Censorship safe.
- Needs to the decentralization anthropology (Figure 6.1).

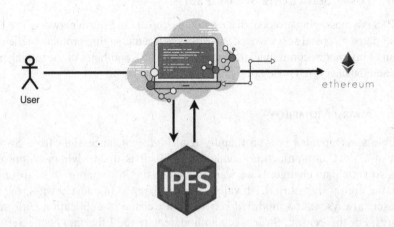

FIGURE 6.1   IPFS workflow.

To further use innovation, FileCoin was proposed as a method for making a decentralized storage system by utilizing the unused storage lying around and boosting clients to be a part of the sharing economy through FileCoins (FIL) by utilizing the IPFS convention. Like other decentralized storage solutions, FileCoin opens up a different economy and perspective around incentives and interest through networking. Individual are rewarded for serving and facilitating content on their space hard drives in remuneration for FIL. It likewise can work with other blockchain networks as a decentralized off-chain storage solution which would be fascinating to see.

6.4 SIA-SIACOIN

Siais adecentralized cloud storage platform that uses a blockchain to facilitate payments. On Sia, you can rent storage, get paid to host files, mine Siacoin, or contribute to the project. Existing cloud storage providers are centralized. Siacoin was developed in September 2013 by DavidVorick and Luke Champine when they were pursuing their undergraduate degree as a computer science student at Rensselaer Polytechnic Institute in Troy, New York. Siacoin is a decentralized file storage system [12]. By these people rent out their unused disk drive space to earn some Siacoin. It uses blockchain technology for file storage. It is using a special kind of blockchain smart contract that provides a way for hosts to get paid for proving that they have stored a file and made it available for end-users over a while. All other payment technologies more optimize the Siacoin network for file storage.

6.4.1 SIACOIN DESIGN FEATURES

Siacoin is written in Go programming language, which is developed by Google. It runs almost in every 64-bit systems. It has a graphical user interface for all end-users, and it also has a command-line version for use in headless environments. Both the interfaces use the same code and the same features and capabilities. It breaks all the files in small blocks.

It is the cryptocurrency, which is used to sell or buy file storage on Siacoin platform. It uses proof-of-work (PoW) algorithm. This Siacoin is created with the proof-of-work mining to stop attacks and to secure the network. It plans to utilize proof-of-burn algorithm for providing more security [13].

6.4.2 FEATURES OF SIA

- **Decentralized**: While Nebulous Labs build up the product, the system is decentralized, implying that no focal authority administers exchanges, record contracts the board or coin issuance.
- **Security using Encryption**: Files put away in Sia are scrambled utilizing the two cryptographic calculations, and just the tenant claims the unscrambling keys.

- **Un-usefulness**: Files are put away over numerous hosts utilizing Reed-Solomon excess calculation, guaranteeing that regardless of whether various hosts go disconnected, the leaseholder will, in any case, approach his records.
- **Accessible**: Being an aggressive commercial center, typical stockpiling costs are much lower than regular distributed storage suppliers, notwithstanding thinking about the redundancy [3].

6.4.3   Future and Improvements

Siafunds are an incredible model, which is worked in Siacoin stage. Ten thousandSiafunds exist; 1,175 Siafunds were sold in May 2014 for an expected $500,000 to help subsidize the improvement of the Siacoin stage. The staying 8,835 Siafunds are claimed by Nebulous, the parent organization of Sia. Thus, 3.9% of all effective stockpiling contract payouts go to the holders of the Siafunds [4] (Figure 6.2).

From the graph shown in Figure 6.2, it is seen that right now, SC has a market estimation of $0.002406 or 0.00000062 BTC (UTC 08:30). It is present market top is as of now getting exchanged at 95,335,714 USD with 24h vol. Worth of $711,249. Rather than its current value, the cost of SC, about one month back, from a similar outline, was seen as at 0.002371 USD. In this way, during this one

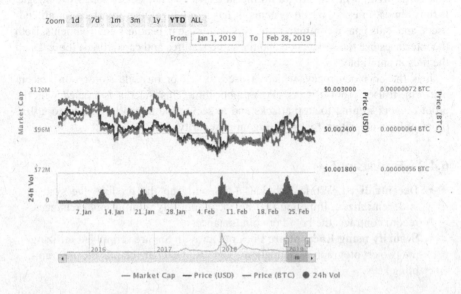

FIGURE 6.2   Siacoin market trends. (From www.coinmarketcam.com.)

month, the cost of SC has climbed by 1.47%. A slight bullish pattern is being seen in this one month with value point 0.002278 USD as an excellent help level [5].

6.5 STORJ PROTOCOL—STORJ NETWORK

At the present stage, the internet is a highly decentralized and appropriate system comprising of numbers of gadgets which are not constrained by a single gathering or different elements of the body. A significant part of the information present is accessible through the internet, is brought together at one place and is put away with a bunch of innovative organizations that have the experience and money to construct large server groups equipped for taking care of this immense quantity of data. A couple of the difficulties looked by server farms or groups are:

* Information leaks
* Number of times its inaccessible on a large scale
* Holding costs
* Increasing size and redesigning rapidly enough to fulfill client need for quicker information and more significant arrangements.

Many decentralized distributed storage is quickly progressing in its development, yet its advancement is dependent upon a specific set of plan limitations, which define the general prerequisites and usage of the system. At the point when configuration in conveyed stockpiling framework, there are numerous parameters to be advanced and developed, for example, speed, limit, byzantine adaptation to internal failure, cost, data transmission, and inactivity. Here a system that scales on a level plane, to exabytes the information stockpiling over the globe. The framework, the Storj network, is a powerful tool that store, encodes, secures, shares, and disperses information to a different number of hubs around the globe for different capacity purpose. Data is put away and served in many ways, deliberately intended to forestall ruptures. To achieve this level, we planned our framework to be measured, comprising of free parts with assignment specific occupations. Here, we incorporated these segments to execute a particular decentralized item in a stockpiling context that isn't just valid, performant, and solid yet additionally significantly more affordable than either on-premise or customary, brought together distributed storage.

6.5.1 DIFFERENT DESIGN CONSTRAINTS

Before planning a framework, it's almost essential to go for its prerequisites. There is a wide range of proposed approaches to structure a chain of the decentralized stockpiling framework. In any of the above case, with the expansion of a couple of prerequisites, the potential structure space shrivels significantly. Our item intensely influences our plan imperatives, and these are efficiently market fit objectives. On cautiously thinking about every necessity, it guarantees that the structure that picks is as all-inclusive, as could be expected under all the circumstances, given the requirements.

6.5.2 Privacy and Security Aspects

The object stored in different storage platforms must guarantee both the protection and security of information, put away paying little heed to whether it is unified or decentralized. Decentralized capacity stages must alleviate an extra layer of multifaceted nature and hazard related to the capacity of information on innately un-trusted hubs. Since decentralized capacity stages can't take a significant number of similar alternate ways server farm-based methodologies can (e.g., firewalls, DMZs, and so forth.), decentralized capacity must be structured from the beginning of the start to finish encryption as well as upgraded security and protection at all degrees of the framework [16].

6.5.3 Concept of Decentralization

Inadvertently, a decentralized application is an administration that has no single administrator. Besides, no single element ought to be exclusively in charge of the expense related to running the administration or have the option to cause an administration intrusion for different clients. One of the principal inspirations for favoring decentralization is to drive down foundation costs for upkeep, utilities, and data transmission. We accept that there are significant underutilized assets at the edge of the system for some smaller administrators. As far as we can tell building decentralized capacity systems, we have discovered a long tail of assets that are unused or underused that could give reasonable and topographically appropriated distributed storage. Possibly, some smaller administrator may approach more affordable power than standard server farms, or another short administrator could contact more affordable cooling. An extensive parcel of these low executive circumstances is not generous enough to run an entire server. For instance, maybe an independent venture or home network attached storage (NAS) administrator has enough abundance of power to run ten hard drives, yet not more. We have discovered that in total, enough small administrator situations exist with the end goal that their blend over the Web comprises significant opportunity and a bit of leeway for more affordable and quicker stockpiling.

6.5.4 Stability and Situation under Device Failure

A capacity stage is futile except if it additionally works as a recovery stage. For any capacity stage to be profitable, it must be mindful so as not to lose the information it was given, even within sight of an assortment of potential disappointments inside the framework. Our framework must store data with high strength and have an insignificant danger of information misfortune. For all gadgets, part disappointment is a certification. Every single hard drive flop after enough wears [6] and servers giving system access to these hard drives will likewise, in the long run, fall flat. System connections may pass on, control disappointments could cause ruin sporadically, and capacity media become untrustworthy after some time. The information must be put away with enough excess to recuperate from individual segment disappointments. Maybe more significantly, no data can be left in an isolated area indefinitely. In such a domain, excess, information upkeep, fix, and substitution of lost repetition must be viewed as unavoidable, and the framework must record for these issues.

6.5.5 LATENCY IN STORJ PROTOCOL—STORJ NETWORK

In the decentralized capacity, frameworks can profit by substantial open doors for parallelism. A portion of these open doors incorporates expanded exchange rates, handling abilities, and large throughput notwithstanding when individual system connections are moderate. In any case, parallelism can't, without anyone else's input, improve dormancy. On the off chance that a unique system connection is used as a feature of activity, its dormancy will be the lower destined for the general action. Subsequently, any circulated framework planned for superior applications should consistently and forcefully improve for low inertness on an individual procedure scale as well as for the framework's whole design.

6.5.6 SIZE OF OBJECTS

We can extensively arrange huge capacity frameworks into two gatherings by standard particle size. To separate between the two collections, we group an "enormous" file as a couple of megabytes or more prominent in size. A database is the favored answer for putting away numerous little snippets of data, though an article store or file framework is perfect for putting away various huge files. The underlying item offering by StorjLabs is intended to work principally as a decentralized article store for bigger files. While future upgrades may empower database-like use cases, object stockpiling is the dominating first use case portrayed in this paper. We made protocol design decisions with the assumption that the vast majority of stored objects will be 4MB or bigger. While littler files are bolstered, they may be all the more expensive to store. This mustn't contrarily affect use cases that require perusing loads of data smaller than a megabyte. Clients can address this with a pressing system by totaling and putting away numerous little data as one large file. The convention supports chasing and gushing, which will enable clients to download little files without requiring full recovery of the totaled article.

6.5.7 CONCEPT OF FRAMEWORK

After having thought about our plan limitations, this part plots the structure of a system composed of just the most fundamental segments. The composition depicts the majority of the layers that must exist to fulfill our requirements. For whatever length of time that our plan imperatives stay steady, this system will, as much as is plausible, depict Storj both now and a long time from now. While there will be some structure opportunity inside the system, this system will hinder the requirement for future-designs altogether, as free parts will have the option to be supplanted without influencing different segments.

6.5.8 Framework Overview

All designs within this framework will do the following things:

- **Store information**: At the point when data is put away with the system, a customer encodes and splits it up into various pieces. The pieces are disseminated to peers over the network. At the point when this happens, metadata is created that contains data on where to find the information once more.
- **Information retrieval**: At the point when information is recovered from the system, the customer will first reference the metadata to distinguish the areas of the recently put away pieces. At that point, the parts will be recovered, and the first information will be reassembled on the customer's nearby machine.
- **Manage information**: At the point when the measure of repetition dips under a specific limit, the necessary information for the missing pieces is recovered and supplanted.
- **Go for Pay per usage**: A unit of significant worth ought to be sent in return for administrations rendered. To improve understandability, we separate the structure into a gathering of eight autonomous segments and afterwards join them to shape the ideal system.

The individual components are:

- Holding nodes
- P2P communication and discovery
- Un-usefulness
- Data about data
- Security using encryption
- Inspection and standard
- Data modification
- Assets transfer

6.5.9 Strong Implementation

To accept the edge work, those are portrayed to be moderately central given plan requirements. In any case, inside the structure, there still stays some opportunity in picking how to actualize every part. In this area, we spread out our underlying execution technique. We expect the subtleties contained inside this segment to change steadily after some time. Be that as it may accept the intricacies sketched out here are suitable and bolster an effective execution of our structure fit for giving profoundly secure, performing, and robust creation evaluation distributed storage. Similarly, as with our past form, we will distribute changes to this solid engineering through our Storj improvement proposal process.

6.6 FILECOIN

Filecoin ($\nrightarrow$) is an open-source, open, cryptographic cash and propelled portion structure proposed to be a blockchain-based agreeable computerized electronic storing and data recovery method. Since the beginning of the decentralization revolution in 2009, plenty of promising activities have come up and changed how we see and live in this world. One of such projects is IPFS which lacks a layer, which can help in its mass selection and thus its ultimate goal to replace HTTP.

Filecoin is a decentralized and distributed network. It allows users to buy and sell unused storage on an open market.

Through expected consensus, the network maintains agreement over the current state of a replicated state machine. This replicated state machine operates the Filecoin storage market. This market facilitates a place to sell and buy storage within the distributed network of Filecoin miners. The exchange also facilitates the needed mechanisms to ensure that the data stored by the system is being stored as promised, without essentiality of the client interaction.

Clients interact with the system by sending messages to the network. These messages are collected and included by miners in blocks. All these messages define a state transition in the state machine. The most straightforward notes say like "move Filecoin from this account under my control to this other account," but even more complex ones describe storage sector commitments, storage deals struck, and proofs of storage.

The Filecoin protocol itself is a set of protocols, including:

- The chain convention to spread the blockchain's information
- The square mining convention to create new squares
- The accord instrument and guidelines for conceding to standard blockchain express all connecting with the state machine and the on-screen characters running on it
- The capacity advertises convention for capacity excavators to offer stockpiling and customers to buy it
- The recovery advertises convention for recovering documents
- The installment channel convention for moving FIL tokens between various regions.

Filecoin network work

Filecoin has three groups of users: user groups, storage miners, and retrieval miners.

- **User groups**: Clients pay for data storage and retrieval. They can find from the available service providers. If the personal information is needed to be stored, they need to encrypt it before submitting to the providers.
- **Storage miners**: It stores the data of the clients for the reward. They decide what quantity area they're willing to order for storage. After the shopper and storage jack have agreed on the deal, the miner is obliged to provide continuous proof that he stores the data. The tests are examined and confirmed for the reliability of the storage jack.
- **Retrieval miners**: It provides client's data at their request. They can obtain information from either shoppers or storage miners. Retrieval miners and shoppers exchange information and coins exploitation micropayments: The info is split into items and shoppers pay a little variety of coins for every piece. Retrieval jack also can work as a storage jack.

Finally, all the full nodes are represented by the network that validates the actions of clients and miners. These nodes count the storage provided, verify the proofs of storage, and repair information faults.

6.6.1 FILECOIN USED AS A BLOCKCHAIN

A blockchain is an open computerized register, which comprises records of exchanges in subsequent request and freely in a progression of connected records, or squares. Rather than relying upon a focal power, it is worked by a system of hubs, each taking an interest in the order and arriving at understanding through agreement conventions.

Though most blockchains utilize proof-of-work mining, Filecoin uses proof-of-storage. The probability that you will mine Filecoin square augmentations with the proportion of limit you give on the framework.

The Filecoin blockchain style is practically equivalent to Ethereum: Filecoin messages square measure generally appreciate Ethereum exchanges, and Filecoin entertainer's square measure comparable to Ethereum contracts.

6.6.2 SIMILARITIES AND DIFFERENCES

All undertakings are comparable in the manner they are attempting to accomplish the decentralized capacity to the groups through a persuading model, which makes it a severe aggressive commercial center. In the specific viewpoints, the isolating elements between them are the arrangement and their agreement computations, so the ones that will lead into the market is the system and the quantity of individuals they get onto their framework which will create eagerness to see.

6.7 CONCLUSION

With any new developing innovation, it's tough for the gatherings to go out on a limb for another decentralized model because there will consistently be a requirement for centralization to control information protection. Regardless of whether it is the smaller to the massive change in the manner substance and document stockpiling is tended to through IPFS or another shared adaptation of Dropbox gets made, it will be genuinely fascinating to perceive what unfurls in the years to come.

REFERENCES

1. Finley, K. "The inventors of the internet are trying to build a truly permanent web". Retrieved from www.wired.com/2016/06/inventors-internet-trying-build-truly-permanent-web/.
2. TechCrunch. Retrieved from https://en.wikipedia.org/wiki/InterPlanetary_File_System#cite_note-ambercase-1.
3. "Sia developers". Retrieved from https://siawiki.tech/about/introduction_to_sia.

4. Robjeiter. (2019). Siacoin (SC). All about cryptocurrency. Retrieved from https://en.bitcoinwiki.org/wiki/Siacoin.

5. Siacoin Is on Its Way to Plunge Deeper into the "Profit Pool". (2019). Retrieved from www.cryptonewsz.com/siacoin-is-on-its-way-to-plunge-deeper-into-the-profit-pool/10153/.

6. "Showdown in the Cloud: Dropbox IPO, Meet the Filecoin ICO". Retrieved from http://observer.com/2017/07/dropbox-ipo-filecoin-ico/.

7. Tepper, F. "Filecoin's ICO opens today for accredited investors after raising $52M from advisers|TechCrunch". Retrieved from https://techcrunch.com/2017/08/10/filecoins-ico-opens-today-for-accredited-investors-after-raising-52m-from-advisers/.

8. Kumar, P., Kumar, D., & Kumar, A. (2019). Customers perception ATM services. *International Journal of Innovative Technology and Exploring Engineering*, 8(7), 2504–2506.

9. Vigna, P. "Latest Hot Digital Coin Offering: $187 Million in One Hour for Filecoin". *Wall Street Journal*. ISSN 0099-9660. Retrieved from www.wsj.com/articles/latest-hot-digital-coin-offering-187-millionin-one-hourfor-filecoin-1502481514.

10. "Investors poured millions into a storage network that doesn't exist yet". Retrieved from *ArsTechnica*. https://arstechnica.com/information-technology/2017/08/investors-poured-millions-into-a-storage-network-that-doesnt-exist/.

11. Kumar, P., & Kumar, A. (2019). Information technology impact on E-commerce business growth. *International Journal of Innovative Technology and Exploring Engineering*, 8(5), 1014–1017.

12. Laurence, T. (2017). *Blockchain*. Hoboken, NJ: For Dummies.

13. Norton, J. (2016). *Blockchain Easiest Ultimate Guide to Understand Blockchain* CreateSpace Independent Publishing Platform.

14. Gates, M. (2017). Blockchain: Ultimate guide to understanding blockchain, bitcoin, cryptocurrencies, smart contracts and the future of money, CreateSpace Independent Publishing Platform.

15. Nagasubramanian, G., Sakthivel, R. K., Patan, R., Gandomi, A. H., Sankayya, M., & Balusamy, B. (2018). Securing e-health records using keyless signature infrastructure blockchain technology in the cloud. *Neural Computing and Applications*, 3, 1–9.

16. Kumar, S. R., & Gayathri, N. (2016). Trust based data transmission mechanism in MANET using sOLSR. In *Annual Convention of the Computer Society of India*, pp. 169–180. Springer, Singapore.

7 Visualizing Bitcoin Using Big Data

Mempool Visualization, Visualization, Peer Visualization, Attack Visual Analysis, High-Resolution Visualization of Bitcoin Systems, Effectiveness

Basetty Mallikarjuna, T. V. Ramana, Suresh Kallam, Rizwan Patan, and R. Manikandan

CONTENTS

7.1   BITCOIN TRANSACTIONS

The advancement of Bitcoin is praised on recent years, also bitcoin referred to as an innovative kind of digital currency. The escalating Bitcoin rate and unstable market in the recent years have prompted substantial awareness from both the financial institutions and technology sectors, making this unfolding payment model one of the most persuasive issues recently. The flourishing of Bitcoin has also organized the evolution of several other cryptocurrencies, including Ethereum, Litecoin, and so on.

Bitcoin, also referred to as a virtual currency, is deposited and interchanged only in digital format. Compared to the conventional form of e-cash used, Bitcoin does not depend on a reliable organization like a government banking sector. On the other hand, it is structured based on an open social mechanism of confidence and on motivated alliance. With the inception of Bitcoin known only to a smaller population, Bitcoin has received significant acceptance as it has found to be useful anywhere, provided it has an internet connection. Due to this, several merchants and business establishments have started accepting the payment made by the customer via Bitcoins. Besides, various stock and trading centers now permit trading Bitcoins when compared to the conventional form of currency.

In addition, Bitcoin is found to be translucent and privacy. The property of translucent is said to be established by following certain level of open standards, i.e., by documenting all transactions via a public ledger. Also, the Bitcoin transaction integrity is in turn assisted by extensively believed cryptographic mechanisms. On the other hand, privacy is said to be attained by linking landlords of Bitcoins with non-transparent cryptographic methods that hide the authentic identities backing them.

7.1.1 BITCOIN AND VISUALIZATION

Acquiring awareness into the opaque datasets produced by contemporary computing and sensing arrangements is still predominantly executed by humans in ownership of domain awareness and requisite mathematical and statistical techniques. Visualization has also been shown to be an efficient means in obtaining awareness into the accessible data. A system of interest that produces huge amount of correlated data and lacks significant organized visualization tools is that of Bitcoin. This Bitcoin cryptocurrency form consists of an approval-less public database to which any user holding a tokenized pseudonymous identity has the possibility to write the protocol-conformant data. Because of the identity is complicated through the utilization of tokenized addresses, the possibility to recognize and segregate malicious behavioral patterns pertaining to the data has resulted in the usage of several financial institutions and contract developers. Managing an inceptive graphical inspection is a competent initial step in the data scrutiny system to explore the systematic characteristics of such reciprocated abnormal behaviors.

7.1.2 FUNDAMENTALS OF BITCOIN

In the recent few years, certain new types of payment nodes have gained an increased attention in various communities for exchanging money. Those payments most are specifically constructed on the basis of digital platforms. The transactions have made the customers to access their monetary benefits more flexibly and all over the world in a very swift manner. Some of the fundamentals of Bitcoin are given in Figure 7.1.

As shown in Figure 7.1, to find out why Bitcoin act as an unspecified payment mode, it is highly required to have a clearly understanding of the protocol design. The format mechanism of Bitcoin is in the form of a network that arrives

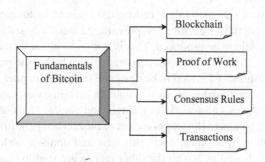

FIGURE 7.1 Bitcoin fundamentals.

at the stature of consensus without possessing any third-party governing body. The entire Bitcoin network is said to be consist of nodes that are closely connected in a peer to peer network. All the nodes engaging the network share a blueprint of a data structure called the blockchain that holds a record of entire transactions in the network. The details of the Bitcoin fundamentals are provided in the following sections.

7.1.2.1 Blockchain

A blockchain [1] refers to a time stamped series of rigid record of data that is controlled by a group of computers not possessed by any single entity or computer. Each of these data blocks are said to be secured and bound to each other via cryptographic techniques.

7.1.2.2 Proof of Work

The main fundamentals and aspects of Bitcoin is the distributed trustless census and hence the elimination of double spending. Hence, during the creation of each block, it must be ensured if a stipulated time was spent into its creation, fortifying that double spending has been made and hence changes to the already created blocks are highly impossible. This is said to be ensured via proof-of-work. In blockchain, this proof-of-work is utilized to ensure transactions and generate new blocks to the existing chain. With proof-of-work, data miners compete against each other to ensure that the entire transactions is said to be completed in a precise manner and hence is said to be rewarded. Here, in this type of network, the users share digital tokens with each other.

With the assistance of a decentralized ledger, all the transactions are said to be collected in the form of blocks. However, utmost care and responsibility are said to be considered while designing the transactions and arranging the blocks accordingly. The utmost care is said to be taken care by the special nodes called the miners and hence the entire process is referred to as the mining. The working policy and methodology involves a complex mathematical puzzle to be solved and a probability to provide the solution precisely and concisely. With this objective, several cryptographic hashes are utilized in the current years possessing random nature and even minor changes that in turn produces completely different hash to be identified by abnormal persons. Figure 7.2 given below shows the schematic view of proof-of-work.

The proof-of-work scheme is designed on the basis of including a nonce to the existing hash value of the succeeding and prevailing block hashes, followed by which yet another hash is said to be produced. This is said to be either lesser or greater than the threshold. Here, the threshold is used to determine whether the proof of work was solved or not. The threshold is said to be designed in such a manner that it is neither found too complex nor found to be very simple. In certain cases, two or more nodes are said to satisfy the proof-of-work for a single and similar block at the same time. This would in turn results in the two branches of blockchain [13]. In this cases, upon obtaining of solution of another block, this new and updated nodes switch to the stronger branch, followed by which the other blocks are said to be eliminated, that again results in a valid single branch.

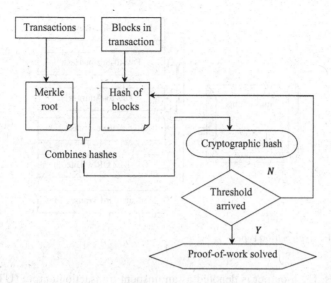

FIGURE 7.2   Schematic view of proof-of-work.

7.1.2.3   Consensus Rules

The nodes in the Bitcoin network is said to validate a block. It also includes all transactions so that durability and privacy of the blockchain is said to be ensured. In specific, the set of rules are said to be consensus rules and these rules are said to be fulfilled by a block. Certain consensus rules include, mining reward, ensuring format of data, double-spending of transaction inputs and signature correctness, and so on.

7.1.2.4   Transactions

A transaction involves the probability for a user to acquire or spend money. Certain basic components of transactions are listed in Figure 7.3.

As shown in the figure, the constituent of transaction includes four basic variants through which the entire transaction is said to be processed. They are input, output, version number, and lock time. Every transaction is said to possess several inputs

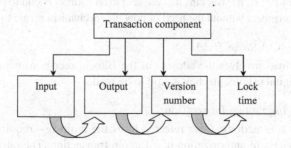

FIGURE 7.3   Component of transaction.

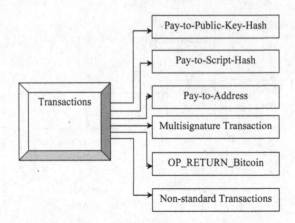

FIGURE 7.4   Transaction types.

and outputs. Each output is denoted as an unspent transaction output (UTXO). Six different transaction types are said to exist (see Figure 7.4).

The transaction types are said to be of six different types. They are Pay-to-Public-Key-Hash (P2PKH), Pay-to-Script-Hash (P2SH), Pay-to-Address (P2A), Multisignature Transaction, OP-RETURN_Bitcoin, and Non-standard Transactions. Each of the six different types of transactions is described in the following sections.

7.1.2.4.1   Pay-to-Public-Key-Hash (P2PKH)

One of the most preliminary forms of performing a transaction in a Bitcoin network is pay-to-public-key-hash. Transactions that are said to be paid to a Bitcoin address consist of a P2PKH scripts. The scripts are said to be resolved by sending the public key and a digital signature. Both this public key and digital signature are said to be created by the corresponding private key.

7.1.2.4.2   Pay-to-Script-Hash (P2SH)

The pay-to-script-hash permits all transactions to be sent to a script hash. The address here starts with 3 instead of a public hash key starting with 1. In order to send the Bitcoins via P2SH, the receiver of the Bitcoin must give a script that should match with the script of the hash value. In addition the data should make the script evaluate to true. Besides, with the usage of P2SH, the Bitcoins that are sent to an address is said to be of a secured form in several manners without the need of how the mechanism is said to be secured.

7.1.2.4.3   Pay-to-Address (P2A)

The pay-to-address involves the address of the Bitcoin receipt to whom the digital payment via digital cryptocurrency is said to be made.

7.1.2.4.4   Multisignature Transaction

Multisignature transactions—also referred to as the multisig—requires more than one key for successful authorization of a Bitcoin transaction. The purpose of using this multisignature transaction is to split the responsibility for holding of the Bitcoins.

The conventional type of transactions usually holds only single-signature transactions. This is because of the reason that transfers here only necessitates one or single signature from the owner with the corresponding Bitcoin address. Despite, single-signature transactions the Bitcoin also supports complex transactions that usually necessitate the signatures of several users or people prior to the transferring of the funds. In that cases the multisignature transactions are used.

7.1.2.4.5   OP_RETURN

In order to perform validation, validation scripts are used in Bitcoin transaction. In turn, the validation scripts are said to be selected from a different palette of predefined functions. This is because of the security reasons only certain functions are allowed to be performed in conventional transaction types. This function is referred to as OP_RETURN.

7.1.2.4.6   Non-standard Transactions

When compared to standard transactions, non-standard transactions as the name implies are first non-standard in nature. Hence, these non-standard transactions are not said to be permitted in the mempool of default-configured nodes. Due to this, these transactions are not said to be transmitted or relayed via the Bitcoin network. In other words, whenever, a data miner looks into the non-standard transactions, only upon the successful pass of validity check, it is transferred to the next stage. Hence, as non-standard transactions are said to be made, they are said to be submitted only to the miner, resulting in systemic inefficiency.

7.1.3   CORE CONCEPTS IN BITCOIN

Four key concepts are involved in Bitcoin. The four key concepts are given in the Figure 7.5.

The detailed description of the four core concepts in Bitcoin are disintermediated, distributed, decentralized and trustless nature and are described in the following sections.

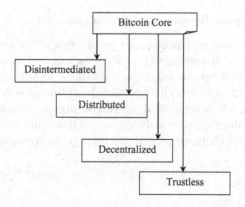

FIGURE 7.5   Bitcoin core concepts.

7.1.3.1   Disintermediated

Whenever, monetary transactions have to be performed between users over internet, there arises a requirement of third party like banks in the banking sectors that manage all the financial transactions of the users. But in case of Bitcoin, transactions are said to be performed in a direct manner, without the aid of the third party over the internet. This type of transaction is said to be taken place in the Bitcoin network. This Bitcoin network, on the other hand and when compared to the third party, does the process of confirmation and verification of transfer of monetary transactions between two users. This concept is referred to as the disintermediated. Besides, the disintermediated refers to the act of removal of the middleman. This disintermediated principle remains the core elements that make blockhain valuable.

7.1.3.2   Distributed

The second core concept related to Bitcoin is that the entire Bitcoin network executes on a network of thousands of computers distributed in nature that shares the work of each other. So, rather than possessing a single centralized computer that handles the entire workload, the work is said to be distributed between multiple computers. Besides, the distributed network is said to be genuine as there is no single point of negligence. Therefore, the work is said to be dispensed covering thousands of computers that are executing and sharing the workload.

7.1.3.3   Decentralized

The third core concept in Bitcoin is the decentralized nature. In other words, it refers to that there remains no intermediate power, no intermediate data repository and no middle management that studies what Bitcoin does. Hence, there remains no intermediate point of failure.

7.1.3.4   Trustless

Finally, trustless remains the fourth core concept behind Bitcoin. This is because that it does not require a third-party. In other words, no bank is said to exist, instead the entire process is said to be performed by distributed trustless consensus. The consensus here refers to that all the nodes agree that a transaction has said to take place [27].

7.1.4   Challenges for Bitcoin and Cryptocurrencies

Blockchain, Bitcoin, and cryptocurrencies are the talk of the town in recent days. There are numerous of disputes as well as about the future of Bitcoin and cryptocurrencies. Some research persons judge that the coming of Bitcoin technique is very doubtful. Some business analysts have predicted that Bitcoin is the next comprehensive digital currency. It is anticipated to bring insurgence in the financial system. On the other hand, there are several disadvantages that could eventually result in its downfall. Some of the challenges for Bitcoin and cryptocurrencies are:

- Volatility
- Ease of utilization
- Universal acceptance
- Prospective for theft

7.1.4.1   Volatility

Bitcoin has been strangely volatile since its establishment. This nature has made Bitcoin exceptionally accepted between speculators who purchase anticipating the price will pursue to increase, but it isn't serving to fire Bitcoin's vogue as a currency.

7.1.4.2   Ease of Utilization

To be honest, it has acquired much easier to purchase, market, and utilize Bitcoin over the formerly several years. But it still isn't user-friendly enough to motivate conventional application. For example, if a person wants to purchase Bitcoin, he or she has to open an account at a Bitcoin exchange, associate a checking account and in several cases delay several days for the transaction to be precise.

7.1.4.3   Universal Acceptance

There are several dealers, specifically online, via which customers pay for transactions in Bitcoin. However, the digital currency still isn't anyplace close to being universally accepted.

7.1.4.4   Prospective for Theft

Security estimates survive that make Bitcoin practically unmanageable to theft. However, taking edge of them necessitates a composite comprehension of how Bitcoin works and hence, would necessitate appreciably more endeavor on the user side. Despite, with online Bitcoin wallets, there's always certain level of possibility that the currency could be theft [18].

7.2   VISUALIZING BITCOIN TRANSACTIONS USING BIG DATA

The blockchain technology became familiar simultaneously with the Bitcoin ecosystem. It furnishes the eccentric chance to document the record of digital transactions and events on the basis of the consensus system that is found to be publicly accessible to all. The publicly accessible transaction data stockpiled in the blockchain gives the research group the potentiality to examine the succeeding and preceding cash flows.

7.2.1   Influencing Visualizing Approaches for Big Data

The elevated price increase [2] of Bitcoins in the last few years has resulted in the increase in the intrigue between laypeople and academic persons. Accordingly, the fields of visualization, security, and privacy became of considerable significance for the investigation and investigation of this virtual cryptocurrency under several facets and situations. Several visualization approaches are said to be divided into three types, based on the utilization and objectives. They are:

- Economic visualization for big data
- Real-time visualization for big data
- Security and analytic-related visualization for big data

The following sections look at these approaches.

7.2.1.1   Economic Visualization for Big Data

From the point of view of economic visualization, the perspective of econo-
mists regarding the Bitcoin payment system is measured from different angle.
From this perspective, the major objective remained in using the Bitcoins, providing solutions to the question of whether it has to be used as a currency or a
commodity. To visualize the economic angle, several graphical patterns were
evolved without any association and exclusively concentrate on widespread com-
pendiums. Several network and conversation analysis were made regarding the
economic visualization for big data and they were able to understand those consequences to Bitcoin. Due to their economic visual representation for big data,
they were accomplished to ascertain an association between discussion sentiment and the prevailing Bitcoin price. Their economic visualizations in connection to the prototype, however, concentrated only on economic states regarding
visualization for big data [14,15].

7.2.1.2   Real-Time Visualization for Big Data

Real-time visualizations for big data of transaction flows are usually published as
web applications. The two major goals for real-time visualization for big data are the
values of the prevailing transaction and frequencies and their association between
each other. The real-time visualizations for big data span from blob visualization
and simple graphical patterns to more complicated world maps and city blocks.
This real-time visualization for big data mostly has a very unforgettable dispens-
ing of the prevailing blockchain transaction. However, the objective information that
can be extracted of those is found to be very less velocity while visualizing the big
data [16,17].

7.2.1.3   Security and Analytic-Related Visualization for Big Data

Finally, the security and analytic-related visualizations for big data has the major
influential factors [11,12,19–26].

7.2.2   Volume and Flow Visualization Patterns

Three types of volume and flow visualization patterns are said to exist. Figure 7.6
shows the types of patterns.

The three types of views are said to exist in the volume and flow visualization
patterns and they are discussed in the following sections.

7.2.2.1   Filter View

The first step in the volume and flow visualization analysis is to filter out elements
with certain attribute levels. For example, it may be of interest to filter out elements
with only one transaction or one-time user or elements that have not been in the
enrollment anymore for a longer period of time. Here, with the aid of a decision tree,
the filtering functionality is said to be applied to filter out the unnecessary elements
based on a decision making process.

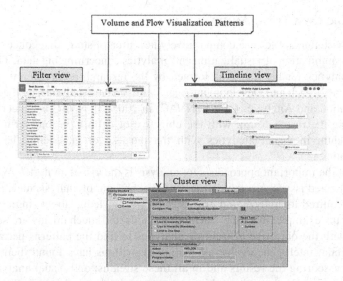

FIGURE 7.6   Different volume and flow visualization patterns.

7.2.2.2   Cluster View

The second step in the volume and flow visualization analysis is to obtain the filtered set of elements. These filtered set of elements are then said to be clustered. This is performed with the objective of clustering the similar elements into a single group. For this specific purpose, the data analysis identifies the number of attributes that are of high relevant and also the total number of clusters. In this manner, the elements with unique higher amounts of Bitcoin transferred are said to be clustered with the objective of making fair comparisons between elements transferring high amounts of Bitcoin.

7.2.2.3   Timeline View

The final and the last step in the volume and flow visualization analysis is the timeline view. This timeline view comprises a number of horizontal timelines and vertical timelines. Here, two axes representing the horizontal and vertical axis are said to be formed using different types of available graphs. The data analyst now performs the task of extracting the temporal distribution of transactions belonging to each cluster. From this, further analyses are said to be made, and final information is said to be obtained in the timeline.

7.2.3   Information Visualizations

The main objective of information visualization is to validate the user to traverse enormous amount of abstract data. A straightforward differentiation and glance via this immense amount of data is impracticable for a human being. Therefore, a visual representation is requisite to obtain a feeling of the organized and the transaction process that are stored in a public manner in the blockchain.

7.2.4   BIG DATA VISUAL ANALYTICS

Big data visual analytics is a comparatively new area of study that concentrates on the tight combination of visualization and analytics concerning big data. The name, visual analytics for big data was formed by the research report Illuminating the Path: The Research and Development Agenda for Visual Analytics, by the National Visualization and Analytics Center (NVAC) in 2005 [3]. NVAC is concerned with sponsoring the research unit with the objective of defining a long-term research memorandum in visual analytics with the purpose of enhancing the analytic potentialities.

One of the important approaches for analysis is the visual analytics. With heavily complicated issues, a combination of higher amount of analysis mechanisms is heavily required to analyze and provide a clear knowledge about single problem. To state, for example, machine learning methods and mechanisms are said to be applied with the objective of training analytics to find the patterns pertaining to specific data; intelligence-value-estimation algorithms have found their place in ranking or scoring the results and in all these streams; and visual analytics have found their place in throughout this entire process to boost these algorithms or to present the results for several interpretation. Some of the applications are shown in Figure 7.7.

As illustrated, having a visual presentation permits a user to observe hidden relationships not identified via algorithmic steps. The inception of big data affects all these analytic methods and elements of the analytic process. Modifying and registering visual analytics to big data issues new advantages and opens new research perspectives.

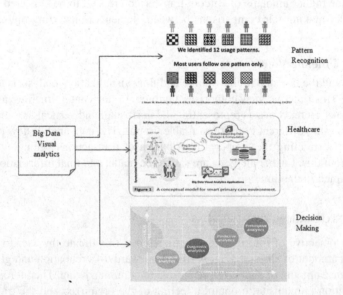

FIGURE 7.7   Sample application scenarios of big data visual analytics.

7.3 VISUALIZING DYNAMIC BITCOIN TRANSACTION PATTERNS USING BIG DATA

For analyzing enormous volume of data, the necessity for storing the data, processing the data and presenting the data are dissimilar than for insignificant more manageable datasets. With the support of specifically designed hardware and software, the utilization of big data has come into existence. Storage system prevailing and pertaining to occupy petabytes specifically higher than that cannot suite the commodity hardware. Hence, specialized type of hardware is said to be required to preprocess these types of petabytes of data. This results in a distributed system where not required a single computer is said to be a part of the three most important layers of data storage, processing, and visualization.

Due to disadvantages of the human aspect and minimum amount of screenspace, it is not possible to visualize large-scale datasets, even if it is found to be highly probable in terms of computation. However, visualizing large datasets without any minimization during the preprocessing phase results in overplotting. This is because of the reason that displaying too many constituents on screen impresses the user and minimizes the visualization efficiency. Though there are several facets to increase the data processing performance, within the visualization pipeline, two types of approaches are present to load and transform data. They are top-down and bottom up.

- **Top-down approaches**: Top-down approaches initiate with the gathering of the elementary data to provide a widespread structure while also circumventing overplotting. This necessitates the gathering of preprocessing step that produces these top-down views. Figure 7.8 shows an example of top-down approach.

  As shown in Figure 7.8, the advantage of the top-down approaches is that the global construction is said to be viewed thin the purview of data and then concentrate on feasibly fascinating features. Having a comprehensive data view, nevertheless, comes at the cost of a preprocessing step that, depending on the data size, to be very costly.
- **Bottom-up approaches**: Bottom-up approaches begin with a subgroup of data and manifest only a slight piece of the comprehensive set. However,

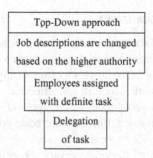

FIGURE 7.8 Schematic view of top-down approach.

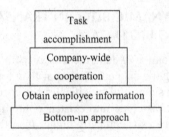

FIGURE 7.9   Schematic view of bottom-up approach.

this necessitates a system analyst to sail within the data to entirely hold worldwide construction of them. Figure 7.9 shows an example of bottom-up approach.

The main advantage behind the bottom-up approach is that the precise data portrayal is obtained from the inception without any preprocessing wait. However, these characteristics consequence in the visualization of only certain portion of data and due to this the user is not conscious of the global constructions.

7.3.1   Mempool Big Data Visualization

The Mempool big data visualization demonstrates the present task and level of relativity as transactions enter the mempool in real time via a repeatedly updated force-directed graph structure. By interconnecting with the Bitcoin network via familiar tarnished addresses, analysis of data is also said to be conducted by recognizing one's own transactions and the responses of the overall networks. Besides, transactions forming independent nature are said to be related to each other in two formats. They are:

- Directly via a prevailing output becoming an input to a new transaction within the timeframe of visualization.
- Indirectly via reusing same cryptographic public keys within an element of transaction.

7.3.2   Peer Big Data Visualization

The aim of this peer big data visualization demonstrates the global scope of the peer-to-peer network pertaining to big data and their visualization. With big data visualization, understanding of network topology is not only prerequisite to provide robustness in the network structure and effective data multiplication but also to analyze which nodes possess higher advantage and which attacks are found to be practical. On the other hand, a Bitcoin Core node cold booting into the peer to peer network commences on a process of network identification via the utilization of hardcoded Domain Naming Services servers. Therefore, with this, it eventually preserves knowledge of up to 2000 peers. The vast majority of peers on the big data network is behind a firewall and hence, it perpetuates their eight outgoing connections solely, while declining all incoming connection requests.

Utilizing big data from blockchain information provides messages pertaining to transactions. It also includes the IP address of the initial peer node that the blockchain information super node is familiar with. Followed by which, the columnar representation is incremented in accordance to the specific IP address to represent the transactional activity. Hence, visualization of this big data in peer environment considerably assists in the lay explanation of a peer-to-peer overlay network, worldwide essence of Bitcoin infrastructure and its task.

7.3.3 BITCOIN MARKET AND BLOCKCHAIN DATA VISUALIZATION

Bitcoin as specified earlier in this chapter as a currency cached and interchanged only in digital format is said to be structured and defined by open standards. It is also said to be controlled by a peer-to-peer network with open membership and is called as the Bitcoin network [4]. The Bitcoins are delegated between users or customers via performing transactions. The transactions being performed are registered in a public ledger. The basics of Bitcoin form the blockchain, also referred to as the public ledger. The recorded transactions maintained in a blockchain are kept at each network node, publicly and permanently unchanged. A blockchain comprises several blocks and several individual blocks consist of one or more transactions. Whenever a new transaction comes into the network, the transactions are said to be validated by each node in the network and finally updated into the blockchain. Figure 7.10 given below shows the simplified structure of blockchain.

As shown in Figure 7.10, the first step remains in the initiation of the transaction. The decision regarding the transaction is acquired from the peer-to-peer network. The peer-to-peer network comprises of all nodes that are presently actively entering into the network. Transactions are then said to be validated and combined with all the other transactions within certain period of time resulting in one block. Upon successful creation of block, the newly created block is then added to the previous blocks. Finally, upon updating of the block, the transaction is said to be completed or accomplished and is not said to be altered any more.

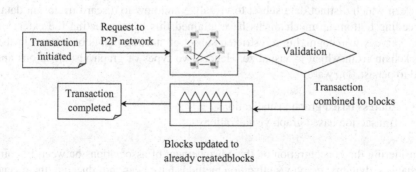

FIGURE 7.10   Overview of blockchain.

The immense evolution of exchanges has encouraged the movement of Bitcoin and ushered the magnification of its value. An immediate look at the Bitcoin blockchain data additionally discloses the superiority of the exchanges, with a larger number of Bitcoins being relocated between exchanges, thus constructing up an enlightened network of transactions. Besides, the analysis of Bitcoin transactions between exchanges with the information of exchange trading methods denotes the best perspective to discern the Bitcoin market.

The Bitcoin public ledger on the other hand called as the blockchain is in the form of a large Directed Acyclic Graph (DAG). The DAG stores all the Bitcoin transactions that have been executed so far. In this DAG, each node in the network is said to form a transaction and it serves as a container of Bitcoins. On the other hand, the edges denote the flow of Bitcoins between transactions. The blockchain here furnishes a wealth of information pertaining to the utilization of Bitcoin. The blockchain are also utilized in tracking and analyzing the Bitcoins flow over time, identifying the patterns of transactions of interest, and so on. From the above said perspective of blockchain data visualization, three major challenges have to be addressed. They are discussed next.

- First, the challenge to be considered is the scalability issue. With the visualization of gigabytes of transaction data, severe scalability issues are said to be arise. Besides, the multifaceted validation of exchange-specific details inflates the increases the problem as well.
- Second, outlining the enlightened trading networks between is significant. This is because of the reason that a wealth of controlling aspects such as their transaction magnitudes, trading densities and topographic localities.
- Third, challenge to be addressed is the visual design. Contributing a comprehensive and complete visual design is demanding. The system being designed has to consider the users' potentiality to acquire abundant information and the scope of restricted user interfaces to exhibit multidimensional data.

7.3.4 BITCOIN EXCHANGES AND DYNAMIC GRAPH VISUALIZATION

The Bitcoin blockchain, with its official ordering of sequences of transactions and correlations between addresses, results in dynamic graph visualization. However, due to the enormous size of transaction, any visualization attempt is strained to settle between which distinct data subset to visualize and how to discard irrelevant data. Preceding bottom-up mechanisms have attained this by obstructing the extent of their surveys to recognizing a restricted subset of starting points of interest in the blockchain from which to visualize. Hence, two types of graph visualizations are said to persist. They are:

- Address based graph visualizations
- Transaction based graph visualizations

Considering the concentration on the advancement of associations between Bitcoin exchanges, dynamic graph visualization methods have been introduced in the recent years. Two surveys [5,6] were conducted that in turn provided a wide viewpoint of dynamic graph visualization. The visual approaches based on the survey were split into three types. The characterized dynamic graph visualization methods are,

7.3.4.1 Animation Approach

Visualization is one of the most robust and dominant methods for representing different types of data, both small and big in nature. Though static data visualizations, specifically are unique ones, and found to be more effective, unforgettable and efficient decision aids than tables and text, animation approach is said to even uplift these edges even further to next height.

7.3.4.2 Timeline Approach

The second approach called as the timeline approach is one of the graphical model of displaying a list of occurrences in either ascending or descending order. Though certain types of timelines use scale to differentiate, certain other types of timelines uses specifically the occurrences of events in a specific order. The main objective of timeline approach remains in communicating the time-related information, either for the specific purpose of analysis or to represent a story.

7.3.4.3 Hybrid Approach

On the other hand, the hybrid approach combines both the animation and timeline. Hence, these types of hybrid approach possess the advantages of both the animation and timeline approach. Contemplating the mental portray necessitated in the animation approaches, the timeline approach specified an edge for the instinctive connection differentiation chore. In specific, this mapping between time and space consisted of two different approaches. They were:

- Node link-based approach
- Matrix-based approach

To that end, the fundamental structure of the dynamic graph visualization with respect to Bitcoin exchanges is summarized as follows:

- **Transactions**: The transactions forming the network are visualized as nodes in an impartial color whose size is stable at the value of the prevailing coinbase reward to provide a sense of scale due to the reason that the size of input and output nodes is found to be changing based on the value. Hence, the transaction node's specific purpose remains only in providing a confined concentration for its analogous inputs and outputs.
- **Input nodes**: The inputs are nodes depicted in an orange color. The size of the input node is said to be proportional to the value of the input node. The input nodes are linked with their containing transaction via an orange edge.
- **Output nodes**: The output nodes are depicted in blue color. The size of the output node is also said to be proportional to its value. The output nodes are linked with their containing transaction via a blue edge. On the other hand, if an output node is said to be referenced as an input node successive transaction within the extent of the visualization, then, the output node is

said to be linked to that specific transaction via an orange input edge. In this way, a chain of spends is said to be formed.

- **Address**: Finally, the addresses are visualized in the form of gray associative edge. This is said to be formed only if more than one input node or output node associates similar address within the extent of visualization.

7.3.5 REAL-TIME VISUALIZATION OF BIG DATA

The visual system of humans of big data has progressed to process the data especially fast and with a high frequency. The real-time visualization of big data is said to be imprudent to leave this prospective unutilized in processing of the information. However, in order to entirely maintain users to acquire apprehend from data, it is also foremost to permit for investigations data analysis via interaction. The solution to this remains in designing a visual analytics that specifically depends on mechanisms from visualization and several other fields, and improves them with data processing and analysis in an automatic manner. Real-time visualization of big data is specifically utilized for random data analysis. However, the visualization part, as well as the automatic machine analysis perspective of visual analytics,is found to be a drawback. This is due to the performance drawbacks in conjunction with big data. Hence, visual analytics of big data is still facing in-depth research. This becomes even more problematic if interactivity is necessary for exploratory approaches.

7.4 EVALUATION OF EFFECTIVENESS OF BITCOIN VISUALIZATION

To evaluate the effectiveness of the visualization of the Bitcoin system, examinations were made on the several visiting groups to data observatory [7,8]. Examinations were conducted among the executive professionals from several organizations, research persons in different domains and also research personnel. From the evaluation, it was found that almost all visitors had the knowledge about Bitcoin and acknowledged it as a currency. Besides, by peer visualization, almost every person identified the mempool visualization as constituting all global transactions, rather than a restricted subset. By explaining the visual transaction format, the users and research persons involved in the data observatory were able to accept the layout of the correlation between transactions far more precisely than the raw data. Besides, the greater number of people involved in the evaluation was then intelligent enough to identify abnormal patterns in the visualization and cross-examine their importance on the basis of the verbal reaction after the original presentation.

- For executives, the discussions inclined concerning about cross-examining the data anonymity to determine the viability of tracing beyond time to decide their emergence. The executives were able to recognize the bulk of established structures, although predominantly were more inclined in the potentiality to relate the other financial transactions.
- For researchers from several aspects and different domain areas, a vast number of observations were made regarding the similarities or likeness in their areas of competence.

- For physicians related to medical and biological fields, to an extent visual comparison were made associated to the network attacks and parasitic organisms. Besides, the structures were easily identified and also block illustrations were made in an efficient manner.
- For persons working in the area of cryptocurrencies, researchers working both internally and externally were benefited. Here, huge visualization size was viewed as a group rather than an individual. Besides, the potentiality to identify an individual transaction from a block was also differentiated effectively.

7.4.1 Benefits of High-Resolution Visualization of Bitcoin Systems

At present transaction rates each block's visualization naturally comprises of a minimum of 5,000 vertices. This is where the prospective of providing high-resolution visualization of Bitcoin systems proves its worth. This is because of the reason that not only is the human vision probable to effortlessly perceive the related patterns of behavior noticeable in the data but one can also physically access the data in detail and perform a rigid analysis of one specific abnormal, while sustaining the background of the entire scenario.

Essentially, performing this high-resolution visualization of Bitcoin systems as part of a team of collaborators has been identified to be most helpful. This is specifically in case when one's head has to be straightforwardly turned to construct correlative inspections beyond multiple blocks concurrently. Besides, the above said reasons, the high-resolution large-scale visualization has found to possess certain amount of supplemental advantages.

- First, by just introducing the entire Bitcoin system to the common public is not found to be a simple chore. By uncovering all of the systems tightly coupled elements at once, providing a thorough clarification and discussing about it via several group discussion can be considerably simplified.
- Second, high-frequency algorithmic behavioral patterns unseen in the opaque data set have now even became directly prominent and distinguishable, significantly quickening further inceptive examination using certain machine learning techniques or pattern recognition algorithms.
- Finally, academic and group discussion in a collaborative manner of the Bitcoin system and its perceived behavioral patterns between both the common public and experts has strengthened the research practitioners in process involving decision making as to where to focus their endeavors.

7.4.2 Human Visual System

Data visualization refers to the graphical representation of abstract information. It is specifically used for two different purposes. They are:

- Sense making also referred to as data analysis
- Communication

Important factors and stories are said to be gathered in our data. Hence, data visualization [9] forms a powerful means to identify and comprehend these stories. Besides, comprehending the stories, the user must be in a stage to present it to others also. The information is said to be abstract in that it narrates the things that are not found to be physical. Besides, the information pertaining to statistical data are also said to be abstract. Some of them are

- Sales volume
- Disease prevalence
- Performance of student

As some of the above instances, even though it doesn't pertain to the physical world, visual appearance is said to appear before us. This transformation of the abstract view to the physical characteristics like, width, height, size, color, shape, and so on is said to be victorious only if the visual perception is said to exist within the user. In other words, in order to visualize the data in an effective manner from the human angle, certain amount of design principles have to be followed. Therefore, data visualization includes both the quantitative aspects and the qualitative aspects. For example, the correlations between users or human beings on certain social networking site such as Facebook, Twitter, Instagram, or between suspected terrorists are said to be drawn with the aid of a node and link visualization.

7.4.3 Expert Analysis

The challenges that big data faces to visual analytics have been exhaustively inspected by a large number of dominant experts. Some of the challenges provided by the expert analysis [10] are listed next.

- In situ analysis (in-memory analysis)
- Interaction and user interfaces
- Large-scale data visualization (visual representation)
- Databases and storage
- Algorithms
- Data movement, data transport, and network infrastructure
- Uncertainty quantification
- Parallelism
- Domain and development libraries, and tools
- Social, community, and government engagements.

Several of the challenges provided in the previous list are found to be precise and are said to be applicable in several areas of big data management, computation, and analytics. However, the challenges that are found to be highly relevant to visual analytics as suggested by the experts are found to be visually applicability and uncertainty. However, visualization of big data specifically requires the construction of an abstract model and visual representations at different levels of abstraction and height. Besides, highly scalable projection of data and techniques involving minimum dimensionality are said to be required to address extreme data scales.

7.4.4 EDUCATIONAL PRIMER OF BITCOIN TO PUBLIC

Several educational primer of Bitcoin to public are said to be available nowadays with the popularity of Bitcoin. Among them, blockchain.com, one of the popular cryptocurrency wallet providers, has declared the inauguration for the launch of a new educational primer called blockchain primers. The educational provider is found to be highly responsible to provide an overview of each crypto asset. The report in narrative included a mixture of both introductory and background information pertaining to Bitcoin and its applicability. It was found to be of highly useful to those users who were found to be less familiar in the Bitcoin area and was also found to be of highly useful with data pertaining to latest market and analysis. Besides, the strengths and weaknesses were also included in addition to the empirical data summarized in the form of graphical representations and tables. An in-depth analysis involving both qualitative and quantitative nature was also included.

REFERENCES

1. Blockchain.info. https://blockchain.info/
2. Historical price data bitoin. https://www.coindesk.com/price/. Accessed: August 22, 2017.
3. Thomas JJ, Cook K. "Illuminating the path: The research and development agenda for visual analytics," *IEEE Computer Society Press*; 2005. ISBN: 0-7695-2323-4.
4. Lu W. bitcoin-tx-graph-visualizer. http://www.npmjs.com/package/bitcoin-tx-graphvisualizer
5. Beck F, Burch M, Diehl S, Weiskopf D. "The state of the art in visualizing dynamic graphs," *EuroVis STAR*, 2, 2014.
6. Beck F, Burch M, Diehl S, Weiskopf D. "A taxonomy and survey of dynamic graph visualization," In *Computer Graphics Forum*, 36, 133–159, Wiley Online Library, 2017.
7. Febretti A, Nishimoto A, Thigpen T. CAVE2. "A hybrid reality environment for immersive simulation and information analysis," In *Proceedings IS&T/SPIE Electronic Imaging, The Engineering Reality of Virtual Reality*, San Francisco, 2013.
8. Leigh J, Johnson A, Renambot L, Peterka T, Jeong B, Sandin DJ, Talandis J, Jagodic R, Nam S, Hur H, Sun Y. "Scalable resolution display walls," *Proceedings of IEEE*, 2012.
9. Rohrer R, Paul CL, Nebesh B. "Visual analytics for big data," *The Next Wave*, 20(4), pp. 1–56, June 2014.
10. Wong PC, Shen H, Johnson CR, Chen C, Ross RB. "The top 10 challenges in extreme-scale visual analytics," *IEEE Computer Graphics and Applications*, 32(4), 63–67, 2012. doi:10.1109/MCG.2012.87.
11. Nagasubramanian G, Sakthivel RK, Patan R, Gandomi AH, Sankayya M, Balusamy, B. Securing e-health records using keyless signature infrastructure blockchain technology in the cloud. *Neural Computing and Applications*, 1–9, 2018.
12. Kumar SR, Gayathri N, Muthuramalingam S, Balamurugan B, Ramesh C, Nallakaruppan MK. Medical big data mining and processing in e-healthcare. In *Internet of Things in Biomedical Engineering*, pp. 323–339. Academic Press, USA.
13. Rana T, Shankar A, Sultan MK, Patan R, Balusamy B. An intelligent approach for UAV and drone privacy security using blockchain methodology. In *2019 9th International Conference on Cloud Computing, Data Science & Engineering (Confluence)*, pp. 162–167. IEEE.
14. Rizwan P, Babu MR, Balamurugan B, Suresh K. Real-time big data computing for internet of things and cyber physical system aided medical devices for better healthcare. In *2018 Majan International Conference (MIC)*, pp. 1–8. IEEE.

15. Patan R, Babu MR. A novel performance aware real-time data handling for big data platforms on Lambda architecture. *International Journal of Computer Aided Engineering and Technology*, 10(4), 418–430.

16. Patan R, Kallam S. Performance improvement IoT applications through multimedia analytics using big data stream computing platforms. In *Exploring the Convergence of Big Data and the Internet of Things*, pp. 200–221. IGI Global, USA.

17. Rizwan P, Babu MR. Performance improvement of data analysis of IoT applications using re-storm in big data stream computing platform. In *International Journal of Engineering Research in Africa*, 22, 141–151. Trans Tech Publications, Switzerland.

18. Khari M, Garg AK, Gandomi AH, Gupta R, Patan R, Balusamy B. Securing data in Internet of Things (IoT) using cryptography and steganography techniques. *IEEE Transactions on Systems, Man, and Cybernetics: Systems*.

19. Karthikeyan S, Patan R, Balamurugan B. Enhancement of security in the Internet of Things (IoT) by using X. 509 authentication mechanism. In *Recent Trends in Communication, Computing, and Electronics*, pp. 217–225. Springer, Singapore.

20. Shankar A, Jaisankar N, Khan MS, Patan R, Balamurugan B. Hybrid model for security-aware cluster head selection in wireless sensor networks. *IET Wireless Sensor Systems*, 9(2), 68–76.

21. Karthikeyan S, Rizwan P, Balamurugan B. Taxonomy of security attacks in DNA computing. In *Advances of DNA Computing in Cryptography*, pp. 118–135. Chapman and Hall/CRC, Florida.

22. Namasudra S, Devi D, Choudhary S, Patan R, Kallam S. Security, privacy, trust, and anonymity. *Advances of DNA Computing in Cryptography*, Vol. 1, pp. 138–150.

23. Poongodi T, Khan MS, Patan R, Gandomi AH, Balusamy B. Robust defense scheme against selective drop attack in wireless Ad Hoc networks. *IEEE Access*, 7, 18409–18419.

24. Krishnamurthi R, Patan R, Gandomi AH. Assistive pointer device for limb impaired people: A novel frontier point method for hand movement recognition. *Future Generation Computer Systems*, 98, 650–659.

25. Selvaraj A, Patan R, Gandomi AH, Deverajan GG, Pushparaj M. Optimal virtual machine selection for anomaly detection using a swarm intelligence approach. *Applied Soft Computing*, 84, 105686.

26. Dhingra P, Gayathri N, Kumar SR, Singanamalla V, Ramesh C, Balamurugan B. Internet of things–based pharmaceutics data analysis. In *Emergence of Pharmaceutical Industry Growth with Industrial IoT Approach*, pp. 85–131. Academic Press, USA.

27. Kumar SR, Gayathri N. Trust based data transmission mechanism in MANET using sOLSR. In *Annual Convention of the Computer Society of India*, pp. 169–180. Springer, Singapore.

8 Blockchain and Distributed Ledger System

Yogesh Sharma, B. Balamurugan, and Firoz Khan

CONTENTS

8.1 INTRODUCTION

Blockchain is a technology that arrived in early 2008 when Satoshi Nakamoto came with the concept of Bitcoin (Nakamoto, 2008) the cryptocurrency. The blockchain technology has grown since then, and the technology is being used in various different sectors of industry as well.

A blockchain is a distributed, decentralized, and a tamper-proof technology. These qualities have made the technology more useful in many of the organizations. A blockchain is a decentralized ledger, or we can say it is like a linked list type of data structure where in nodes are connected with a pointer. Similarly, a blockchain is a chain of blocks connected with each other with the current one referring to previous ones. Each node or a user of the blockchain can create a block and add it in a blockchain. The users of the blockchain network are unaware of the transactions taking place in the network. The blockchain network is a tamper-evident network that means once any transaction or information is added on to a blockchain, that information can never be changed; if any node tries to change the information, other nodes gets the information that there is a change in the transactions and that block will get rejected from the chain of blocks. For this reason, many organization and industries are moving their existing network to the blockchain network, as the blockchain network provides better security, better privacy, and above all transparency among the users.

But the organizations and industries need to keep in mind the fundamental aspects related for implementation of blockchain technology, as it is not easy to export an existing database to a new platform. Consider a situation where an organization has moved on to the blockchain network and then some modifications need to be done in the database. Blockchain technology and a network based on blockchain technology is tamper proof so changes would be very difficult as it would require changes to be made from the starting block.

Since blockchain technology is a decentralized network, that means there is no single owner or central authority of the network, and each node can directly communicate with other nodes. Thus, blockchain can be very beneficial in business organization where the transfer of money is involved from one party to another party without the involvement of a middle man.

However, blockchain technology is still in its growing phase, and organizations should investigate how the industry-based problems can be fitted into blockchain technology before moving onto blockchain network.

8.2 A BLOCKCHAIN PROCESS

A blockchain process is based on the consensus mechanism for a transaction to be verified and valid. A blockchain network consists of various nodes connected in a distributed and decentralized fashion. Any node that needs to perform a transaction has to write the transaction in a block. A block can be thought of a container like a data structure that can consists of about 500 transactions on an average; however, the size of a block can be up to 1 MB (Madeira, 2016). The node performs a trans-action in a block and sends to other nodes, which are connected in a distributed network. Since the blockchain network is decentralized, the block of transaction created by the node, hence the block is being broadcasted to all the nodes present in the network. The blockchain network is a consensus-based network meaning all the nodes present has to agree on transaction made based on consensus algorithm. Consensus algorithm is a process by which multiple nodes present in a distributed network agreed upon a decision (Margaret Rouse, n.d.). The consensus algorithm provides reliability to a distributed network. There are many consensus algorithms like proof of work, proof of stake, Byzantine fault tolerance, and many others. Thus, as shown in Figure 8.1, a block created is added to the chain of the blocks already

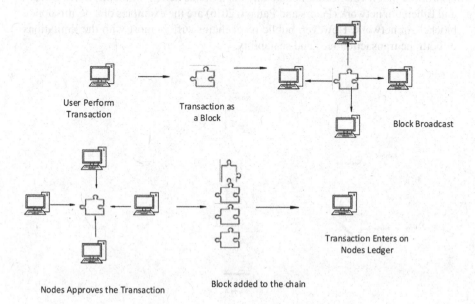

User Perform
Transaction

Transaction as
a Block

Block Broadcast

Nodes Approves the Transaction

Block added to the chain

Transaction Enters on
Nodes Ledger

FIGURE 8.1 Blockchain process.

present in the blockchain once it is verified and validated by all the nodes present in the network as shown in the figure. The newly added block will always refer the previous block in the chain, which will make the chain more secure. Once the block is added onto the chain of blocks, the receiver node can now update the ledger with the new information.

8.3   TYPES OF BLOCKCHAIN

A blockchain network can be classified into three categories based on the transaction processing and the access of data in the network. These classifications can be of great use when an organization thought of either moving their existing database on to a new blockchain network or creating a new blockchain network. The blockchain network can be of three types—public, private, and a consortium type of blockchain. All the three type of blockchain network places important role in any organization and depending upon their functionality an organization can go for any of these networks.

8.3.1   PUBLIC BLOCKCHAIN NETWORK

A public blockchain also termed as a permissionless blockchain. As the name says permissionless, there is no need of any permission to join the network. Any one is allowed to join the network. The public network is open to everyone anyone who wants to read the transactions made on the block, anyone can make legitimate changes on to the network, and anyone is allowed to add a new block of transaction on to the existing chain as shown in Figure 8.2. The biggest advantage of a public network is that it can accommodate any anonymous actors on the network. Bitcoin and Ethereum network (Peters and Panayi, 2016) are the examples of a permissioned blockchain network. However, public blockchains suffers most with the limitations of both the transaction fees and scalability.

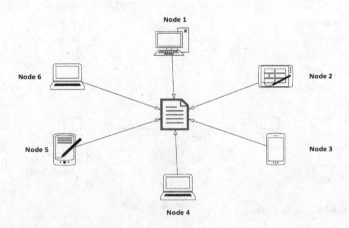

FIGURE 8.2   Public blockchain.

8.3.2   PRIVATE BLOCKCHAIN NETWORK

A private blockchain is also known as a permissioned blockchain. Unlike a public blockchain network, a private type of blockchain is a network has some restriction in it and is not open for any anonymous users. A private or a permissioned blockchain consists of limited users and only authorized users who have access to the network (Figure 8.3). In a private blockchain network the creator of the network will be the administrator or a validator who approves or disapprove any transactions made by the node of the network. In a private blockchain, only the administrator has the access to the transactions and the information present in the network. It is the duty of the administrator to maintain the network and has the authority to change or modify the rules of the network. Hyper Ledger, Ripple (Peters and Panayi, 2016) are the examples of a permissioned blockchain.

8.3.3   CONSORTIUM BLOCKCHAIN

A consortium blockchain can be considered as a semi-private blockchain. In a private blockchain the members of the blockchain are connected in a permissioned environment. The private blockchain is owned by a single company or an industry which is more specifically described as a centralized system but with strong cryptographic methods attached.

A consortium blockchain has the same benefits which a private blockchain provide but in a consortium blockchain the ownership is not in control of a single company or a person rather operates under the leadership of group (Thompson, 2018).

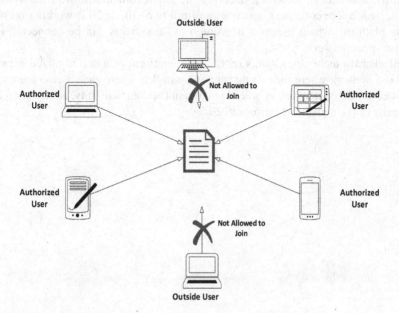

FIGURE 8.3   Private blockchain.

8.4   MAJOR COMPONENTS OF A BLOCKCHAIN NETWORK

1. Peer-to-peer network
2. Distributed ledger
3. Cryptography
4. Consensus algorithm
5. Smart contract

8.4.1   PEER-TO-PEER NETWORK

A peer-to-peer (P2P) network is a self-organizing arrangement of equivalent and self-governing elements that works for the use of shared, distributed resources in a network, which avoids central authority (Ralf Steinmetz, n.d.). In a client-server network one system will be the server and other systems will work as the client systems. In a client server network, clients are always controlled by the server system on the network. In a peer-to-peer network all the systems are connected to each other with same capabilities hence termed as "peer". In this type of network file can be shared with each peer directly and there is no need of a central server that means there is no single system, which will be a client or a server system.

In peer-to-peer network (Figure 8.4) all the systems connected can act as a server and as a client as well. Once a peer is connected to the network, P2P software allows a peer to search for the relevant files on other system. Now the question arises how many systems can be there in a peer-to-peer network, the answer is as such there is no limit on the number of systems to be connected in a peer-to-peer network but more than a limit may cause headache in terms of performance, security, and access. However, according to Microsoft, the limit of the systems connected in a peer-to-peer network can limit to 5, 10, or 20 if working on windows platform, which means a maximum of 20 systems can be connected to a peer-to-peer network.

Blockchain technology works on the same platform of a decentralized network that is a network where there is no central controller, where each user is connected to a decentralized network in order to maintain the confidentiality, authenticity, and integrity of the network (Martinovic, 2017).

FIGURE 8.4   Peer-to-peer network.

Confidentiality is nothing but keeping the records or information accessible to only those individuals who are authorized to access it. Authentication methods can be used to in order to establish the authorization. Hence there are various methods and approaches which can be used to provide confidentiality, which can be either physical protection or some mathematical algorithms that changes the data to a normal user. Thus, the encryption is the most common way of achieving confidentiality where the data or information is converted to a corresponding cipher text which protects the data from unauthorized persons and only authorized person will be allowed to decrypt the data.

The integrity of data ensures that an unapproved person has not changed the data. The manipulation of the data or information can be achieved by simple insertion, deletion, or substitution methods. Now the purpose of integrity is to detect if there is any change in the information and that manipulated data cannot be accepted. There are various cryptography methods by which we can achieve the property of integrity; the most common method is hash function. These functions provide a strong method which provides a compressed representation of the data called as hashvalue and it became practically impossible for a hacker to find the actual value of the input and hence not possible to change or manipulate the information.

Authentication is the process that can be applied to both the user of the network and the origin of the data. The authentication process can be used to identify user when entering into the communication network. There are various methods and protocols useful in claim of identity of the user; digital signature, biometrics, username password are few of them. Similar to this are methods and protocols that can be used to verify the origin of the data, the content of the data, time of the data collected, etc. In blockchain technology the process of authentication is very important as any transaction to be made by the user of blockchain has to go through the process of authentication also it unable the participants of the network to read and write into blockchain, but reading and writing does not mean they can change the rules of the network, especially in case of the private blockchain as only the owner of the blockchain can change the rules.

8.4.2 DISTRIBUTED LEDGER

Before moving on to the concept of distributed ledger we first need to understand the concept behind distributed system on which the distributed ledger technology is based. A distributed network can be thought of a network in which two or more nodes are working together on some information in a coordinated fashion and they produce the same output. A distributed system is made in such a way that whatever output is produced will look to each user a single logical platform.

But the major challenge with a distributed system is in the coordination of the nodes they are working and the fault tolerance. In a distributed system any participating node can continue to perform with the same intensity and obtain the desired output even if any one of the nodes fails or if any link to the network fails.

Thus, on the basis of the concept of distributed system, came the concept of distributed ledger. These distributed ledgers are fast and decentralized and above all are cryptographically secured. We can now define a distributed ledger as a ledger in which all the transactions, information, and records are recorded from different nodes from multiple locations and the data can be shared and synchronized across distributed network thus eliminating the need of a central authority. All the information and records are cryptographically secured and can be accessible only by the keys with cryptographic signatures. With the distributed ledger any changes made in the ledger or the document will get reflected across all the participating nodes in few minutes or seconds and the involved parties can check in their ledger about the changes made and use the updated record.

Since it becomes sometime inefficient to be constantly updating and synchronizing the data across many centralized databases, thus by putting the databases on a shared ledger it would become easier to access and view the updated database on demand when needed. A blockchain technology is one form of distributed ledger technology (DLT) but it is not necessary that a distributed ledger system employs a chain of blocks in order to successfully provide a secure distributed ledger.

There are two main classes of distributed ledger: public ledger and permissioned ledger.

1. Public ledger: The public ledger is maintained by the public nodes and is accessible to anyone. Bitcoin is a well-known example of public blockchain where anyone can join the network, anyone can read the network, anyone can make the legitimate changes on the network, and anyone can write a new block on the chain.
2. Permissioned ledger: In a permissioned blockchain the creators of the blockchain network will act as an administrator/validator who will approve or disapprove the transactions of a permissioned ledger.

8.4.3 CRYPTOGRAPHY

Cryptography is a technique of writing a document into a secret code (Sharma, 2017a). The process of cryptography is necessary when the communication is taking place from an untrusted network, particularly an internet. Cryptography is he process in which a plain text is converted into a cipher text and vice versa. This cipher text is the secretly coded output of the plain text. Thus, the coded message is stored or transmitted in either form so that only authorized person can use the data. The cryptography process not only saves the information from unauthorized users the process saves the document from unintentional modification also.

There are basic five primary functions of cryptography:

1. Privacy and confidentiality: This function ensures that only the intended recipients can read the message and no one else.
2. Authentication: This function of cryptography deals with the process by which a user can prove his/her identity.

3. Integrity: This function assures that the message received by the recipient has not been changed in any manner that was sent from the sender.
4. Non-repudiation: This function of cryptography is used when there is a need to prove that only the sender has sent the message.
5. Key exchange: With this method, the crypto keys are exchanged between the sender and the receiver.

The hash functions are the cryptographic functions created to keep the data secure. The hashing process is an irreversible process unlike other cryptography techniques that is in hashing. When the output or the hash value is obtained it become virtually impossible to reach back to its original value. Thus, the hash functions are also called as one-way functions. This feature of hashing makes the hash algorithms securer as compared to the other cryptographic algorithms. The hash functions are also called as compressor functions as the output obtained through the hashing process is a fixed length output whatever be the length of the input. Thus, a hash values plays an important role in a blockchain technology in keeping the data/transaction secure and tamper proof.

8.4.4 MERKEL TREE

Another important component of blockchain technology is Merkel hash tree or simply a Merkel tree. A Merkel tree, also known as hash tree, was built by Ralph Merkel and patented in 1979. A Merkel tree is also known as binary hash tree which can be used in cryptography. A Merkle tree is the data structure that consists of cryptographic hash values. A Merkle tree can be built by repeatedly hashing the pair of the child nodes until we get a node with only one hash value; this node at the top is called the Merkel root.

The leaf node of the Merkle tree consists of the data; a data can be a file. The data value in the leaf nodes are not the hash values and cannot be considered as part of the Merkle tree. The data values are hashed using SHA-256 algorithm and the resulting hash value are stored on the leaf node of the Merkle tree. Now again the child node at the leaf node are appended and hashed together and merge into a parent node. This process is continued until we receive a single node at the top of the tree and this node is called as Merkel root node shown in Figure 8.5. Now this kind of security is applied in a blockchain network where every transaction is secured with a hash value and every new transaction hash value is derived from the hash value of the previous transactions. So, if there is any change in the data or data is tempered by some unauthorized entity, the root hash value will get change. This will inform the parties of the blockchain regarding the change in the data. Thus, blockchain provide a securer network for every organization and industry.

8.4.5 SMART CONTRACT

When we talk about transactions in a blockchain network, a transaction can be any type it may be a transfer of documents from one party to another one or it could be transfer of money from one person to another, from seller to buyer, be it

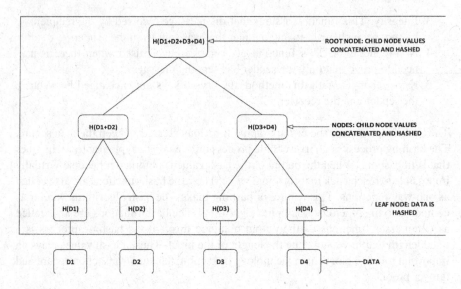

FIGURE 8.5   Merkel tree.

a home seller or buyer or food product seller and buyer. A blockchain technology plays an important role when the transaction between two parties is money based thus eliminating the middle man between seller and buyer. For this purpose, one of the key elements of blockchain used is called as smart contract a concept given by Nick (1996). A smart contract is a small computer code which executes automatically and imposed when a transaction completes between the two parties. A smart contract is a set of rules or a clause between two parties which agrees and governs the relationship. A smart contract act as a contract or an agreement between the parties, now whether two parties agrees or not depends on the parties but the use of smart contract will definitely eliminate the third party. Smart contracts are small computer programs on the blockchain network. The smart contract are self-executing codes that executes automatically when an input value is given to the smart contract the corresponding function in the program code of smart contract gets executed. They are self-verifying codes that are written either in solidity or in python and gets integrated in the blockchain network. Now since the blockchain network is a distributed network this means if a smart contract is added or integrated with the blockchain network every participant will have a copy of the smart contract. Thus, each node in the network will have the history of all the smart contracts, history of all the transactions and the current state of all the smart contracts. The smart contracts are very useful in a private-type blockchain network so that all the nodes are loaded with the agreement, and whenever there is requirement of transaction, the preloaded smart contract can be used at the same time.

Figure 8.6 is the basic structure of smart contract (Arshdeep Bahga, 2016), which consists of value, address, state, and the function.

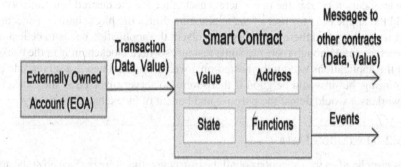

FIGURE 8.6 Basic structure of smart contract. (From Arshdeep Bahga, V.K., *J. Softw. Eng. Appl.*, 9, 533, 2016.)

8.5 LAYERS OF BLOCKCHAIN

The blockchain technology is based on a layered approach Figure 8.7. The blockchain technology is decomposed into several layers that will in turn help in better understanding of security and the design of the blockchain. There are few layers for the blockchain technology discussed in the following sections.

8.5.1 APPLICATION LAYER

As we know a blockchain technology is a tamperproof, decentralized, and shared ledger technology, so there can be multiple applications that can build on the basis of these features of the blockchain. Some applications built in application layer can interface with the other layers; therefore, the application layer is on the top of this layer suit.

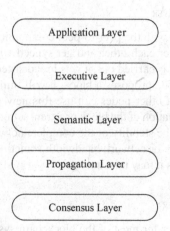

FIGURE 8.7 Blockchain layers.

The application layer is the one where a user can code the desired functionality and build the application for user of the application. Since the blockchain is a decentralized technology and there is no server involved, the application needs to be installed on each node. Although there are some instances where blockchain is in the backend and the applications need to be hosted on a web server and needs a server-side programming, but it would be good if there were no server involved in the blockchain network as it would defeat the purpose and benefit of blockchain technology.

8.5.2 EXECUTION LAYER

This layer handles the executions of all the instruction that were performed at the application layer for all the nodes present on the blockchain network. The set of instruction could range from simple ones to multiple instructions. For example, smart contract is also small code that needs to be executed when some funds need to be transferred form one person to another person. Now if one application is present on all the nodes of the blockchain network the code has to be executed independently on all the nodes. In order to avoid the inconsistencies in the output, the execution of code on a set of input should always produce the same output for all the nodes present on the blockchain.

8.5.3 SEMANTIC LAYER

This layer also called as logical layer of blockchain layer suit. This layer deals in validation of the transactions performed in the blockchain network and also validating the blocks being generated in the network. When a transaction comes up from a node, the set of instruction are executed on the execution layer and gets validated on the semantic layer. Semantic layer is also responsible for the linking of the blocks created in the network. As we already know each block in the blockchain contains the hash of the previous block except the Genesis block. This linking of block needs to be defined on this layer.

8.5.4 PROPAGATION LAYER

A propagation layer deals with the peer-to-peer communications between the nodes that allow them to discover each other and get synced with another node in a network. When a transaction is carried out, it gets broadcasted to all other nodes in the network. Also, when a node proposes a block, it will immediately get broadcast in the entire network so that other nodes can use this newly created block and work upon it. Hence, the propagation of the block or a transaction in the network is defined in this layer and ensures the stability of the complete network. However, depending upon the network capacity or network bandwidth sometimes the propagation could occur instantly sometimes it may take a longer.

8.5.5 CONSENSUS LAYER

This layer is the base layer for most of the blockchain systems. This main purpose of this layer is to make sure that all the nodes must get agree on a common state of the shared ledger. The layer also deals with the safety and security of the blockchain. There are many consensus algorithms which can be applied for generation of crypto-currencies like Bitcoin and Ethereum, they uses proof-of-work mechanism to select a node randomly out of various nodes present on the network that can propose a new block. Once a new block is created, the block is propagated to all the other nodes to check if the new block is valid or not with the transactions in it and based on the consensus from all other nodes the new block gets added on to the blockchain.

8.6   BLOCKCHAIN USE CASES

The blockchain technology has seen major growth in various sectors. The technology is being used in different major organizations with multiple purposes. The technology is not just limited to cryptocurrency but also the performance of the technology. The technology provides high level of security and privacy to the network. The networks based on blockchain technology are improving the growth and quality of the business process. There are many sectors that are already using the blockchain technology for improving of their organizations. A few of the sectors are banking and insurance, healthcare, supply chain, agriculture, real estate, and many more.

8.6.1   BANKING AND INSURANCE

Today, the banking or the financial industry (Marr, 2017) is facing major issues regarding fraud either online or offline on a centralized server. The personnels involved in detecting the fraud are different at various stages, which may cause a delay in resolving the fraud in real time. Thus, banking sector are in the process of finding new ways to enhance the customer service which will provide faster transaction process in a more secure manner and ensuring the transparency in the operations involved. Tejal Shah (2018) has talked about the fit assessment framework in order to understand if the blockchain is suitable for a particular use case or a process. As we know, it is not easy to shift a process from a database-based network to a blockchain. So, the key points framework may be useful if it would be beneficial to move on to the blockchain network.

Since a blockchain is based on the distributed ledger technology, it would be ideal in banking sector as it would reduce the cost involved and maintain the transparency between the customers and bank. The transactions will become more secure and would have increased privacy due to the hash values used in each block of record. It also makes the records tamperproof, thus making the records immutable and irreversible. The feature of smart contract in the blockchain technology also plays an important role by eliminating the middlemen and codifying the process of transaction. All these steps might improve the banking and the financial systems and reduce the fraud involved.

Similarly, in the insurance industry keeping the network on blockchain will be a benefit for the insurance industry. A person taking insurance from the insurance company need to create a block on the chain with medical information and history involved. Now when a person puts in the information because of the tamperproof nature of the blockchain, the person will not be able to change or modify the medical information after it's put in,and the payment done can be executed by the smart contract.

8.6.2 HEALTHCARE INDUSTRY

The healthcare industry is growing day by day and the medical information of patients (lab reports, X-rays, MRIs, previous medical history, and so on) is growing at a greater rate. According to (www.ibef.org/industry/healthcare-india.aspx, 2019) the healthcare market might increase to many folds to Rs. 8.6 trillion (US$ 133.44 billion) by 2022. It becomes difficult for the patient to carry their health records from one hospital to another or from one doctor to another one. Patient records are generally stored on the hospitals database, so there is not security with the medical record stored in a database (Kumar et al., 2019). A patient's medical record can be misused by some other person. So, there was a need for such a network where these electronic medical records could be stored.

Blockchain provides such a network where a patient can store the electronic health record (Dhingra et al., 2019). A blockchain network will provide more security and privacy to the information stored on the network. Once the medical record is stored on the blockchain it becomes very difficult for someone other than the patient to access the record because the records are digitally secured, and with every record a hash value is attached. Thus, if any unauthorized person tried to modify the record of the hash value will get change and the patient will know about the record tampering.

The patient can retrieve his/her medical record anywhere with his/her smartphone or a computer and can also check who is using medical record and for what purpose and can reject the permission to access the document.

The blockchain technology can also be useful in pharmaceuticals industry where the technology can prevent the counterfeit of medicines Mettler (2016) by tracking the medicine from manufacturer to the customer.

8.6.3 SUPPLY CHAIN

According to Consulting (2019) the market size of global supply chain analytics is expected to be around USD 10.7 billion by 2026; this market is expected to grow with 15.5% CAGR during the forecast time period. The supply chain market involves almost all the product of daily usage. There are some barriers associated with the supply chain management (Fawcett et al., 2008) as the persons involved in the supply chain lacks in trust and has poor synchronization between the partners associated in the supply chain. (Jain and Bhatnagar, 2012; Vedpal, Vipul Jain, and Naresh Bhatnagar, 2012) gave the concepts of 4Cs, which are cooperation, coordination, collaboration, and co-opetition. These four Cs play an important role for a successful supply chain management as the cooperation between the parties involved is a must in supply chain, whereas there has to be proper coordination between the supplier and the manufacturer and should be working together. The blockchain use case for the supply chain could be a public type of blockchain where any supplier can join the supply chain network, or it could be a private type of blockchain where only pre-registered or authenticated sellers and manufactures are there in the chain.

When the supply chain network moves onto the blockchain network it is beneficial for the end user as the customer gets the knowledge of original supplier and manufacturer. Thus, a level of trust is created between the creator and consumer. The consumer will also get the update of date of manufacturing and the correct price of the product. With the help of smart contract, the payment can be directly made from consumer to manufacturer and from manufacturer to supplier eliminating the middle person in between.

The use case for supply chain includes multiple supply chain be it a retail chain, a garment supply chain or a food product supply chain. Although a blockchain network may provide many benefits to an industry, but it is not necessary that all these networks can be moved on to blockchain network. The feasibility of the network both with the cost and time should be checked before implementing.

8.7 CHALLENGES WITH IMPLEMENTING A BLOCKCHAIN

A blockchain technology has some great potential to be included into business processes and can benefit the business process management in an innovative manner. A blockchain-based network could be a secured network and could provide full privacy to the network, but at the end of the day the company needs to generate revenue. So, from an individual organizations perspective there are few challenges in adoption of the technology. We all know that the technology is still in its early stage, and there are some limitations and some confusion around certain areas such as system performance, scalability, cost etc., and so on. There is a need to understand business goals and try to align those goals with the choice of technology like a public or private or a consortium type of blockchain to be used and type of consensus mechanism to be used will have a profound impact of the business of an organization. The greatest challenges with adoption of blockchain technology will be industry standards, but what is encouraging is a lot of industries are using or in the near future are planning to move on this decentralized, immutable technology.

The problem with adopting the blockchain technology is the same as moving the existing system from one technology to another different technology. Any industry working on a platform and suddenly moving on to another platform is very challenging for any industry.

Another challenge could be inconsistency with the partner of the technology. Some partners are a little more front footed and ready to invest their money, time and resources in the blockchain technology whereas they could be some participants who are more distracted to some other consideration or resources to contribute. There could an issue in implementing blockchain technology as the technology deployed may be legal in one country and not legal in another county; therefore, it could be problem in interorganization business so the technology has to be global in nature.

Scalability is one of the major concerns of blockchain technology, when the application of payment is talked about. According to Papermaster (2018), a visa process may take up to 1,667 transactions per second whereas Bitcoin takes 3 to 4 transaction and Ethereum takes about 15 transactions per second. The issue of scalability may decrease the speed of transactions and might affect directly or indirectly the business process and the parties involved in the blockchain network.

8.8 DISTRIBUTED LEDGER TECHNOLOGY

A distributed ledger technology, also known as DLT, is a technology based on a peer-to-peer network which is very secure. A distributed ledger is also known as a shared ledger or distributed ledger technology. A distributed ledger is a kind of database of replicated, shared data that is geographically spread across several numbers of nodes, sites, countries, or institutions. There is no central authority present in a distributed ledger system and each node present updates their own database independently. A distributed ledger system works on a peer-to-peer network with some consensus algorithm that ensures the replication process between the nodes. A distributed ledger system can be categorized as either a permissioned or permissionless ledger depending upon whether the users of the distributed ledgers can be accessed by anyone or only some desired participating node can access the ledger. The technology came into existence with the concept of cryptocurrencies like Bitcoin but the technology has moved three-fold and now the technology is being used in many use cases and many organizations are investing huge amounts in distributed ledger technology as it is cost saving measure and, in some way, or other reduce the risk involved in their operations.

8.8.1 EVOLUTION OF DISTRIBUTED LEDGER TECHNOLOGY

Earlier in the banking system all the records used to be maintained manually in registers and on paper. With the invention of computer in around 1980s and 1990s, the records were shifted into some database that was controlled by a single server or a centralized server also called as centralized ledger system. But the problem with this centralized ledger system was that tit was controlled by a single authority, and a single point failure may disrupt the complete network.

The concept of distributed ledger arrived in late 1982 with Byzantine general's problem theorized by Lamport and Shostak (1982), whereas the concept of blockchain came into existence in 1991 by Stuart Haber (1991) and Dave Bayer (1993). The real existence of blockchain technology came with the concept of cryptocurrencies like Bitcoin in 2008 by Satoshi Nakamoto. A distributed ledger technology is based on the trust between the peers connected in the network. A distributed ledger technology adds the trust, transparency, security, and traceability to a network (Nabil El Ioini, 2018). The distributed ledger technology is a decentralized technology, i.e., without any central authority. The DLT has evolved since then, and many blockchain use cases are following the distributed ledger technology.

8.8.2 DIFFERENT TYPES OF DLTs

There are different types of networks that use distributed ledger technology. A DLT is a kind of digital ledger/database that is held by each node present in the network. The ledger or database is distributed evenly to all the participants, and any updates at one node can be automatically done on all of the other ledgers; therefore, in this type of network there is no need of the central authority for updating rest of the ledgers. Different types of DLTs have different approaches to the agreement in order to store the record on to the ledger.

There are five types of distributed ledger technology: blockchain, DAG (directed acyclic graph), HashGraph, HoloChain, and Tempo (Radix).

8.8.2.1 Blockchain

The blockchain technology is one type of distributed ledger technology that has started with the introduction of Bitcoin by Satoshi Nakamoto in 2008. A blockchain is a distributed and decentralized network that is secure and provides a good level of privacy. In a blockchain the records are maintained in a chain of blocks. This chain of block can be of any length depending upon the transactions made on the blockchain. The blockchain is a kind of database storing the records in a chain of blocks. The chain of blocks is a kind of digital information stored in the database called as blockchain. The chain of blocks in the blockchain can be thought of nodes connected in a linked list in the database, where all nodes are connected with each other with a pointer. Similarly, the blocks in the chain are connected with a unique ID attached to each block called a hash. A hash value is always a unique number and cannot be changed or modified as the hash value once created cannot be tracked back to the input value. This hash value is responsible to differentiate between the blocks generated.

8.8.2.1.1 How a Blockchain Process Works

So how does a blockchain process works? There are four main stages of how a block gets added onto a blockchain.

In the initial stage the participants connected on the network need s to perform a transaction it could be a financial transaction.

The transaction created reaches to all the participants connected in the network. Now it's the duty of the participants to verify the transaction created. The process of verifying the transactions is different for every network and depends on the nodes connected. The nodes have to arrive on an agreement that allows a transaction to take place. This agreement is known as a consensus algorithm. After the consensus is made and if the majority of the nodes verify that the transaction is true, the transaction is stored in the block.

Once the transaction is stored in the block, all the information regarding the transaction like date, time, amount, and digital signature gets stored in the block. After this process a hash value is generated for the block, which also gets added with the block. This first block created is called a genesis block. The same process gets repeated for every new transaction, and the blocks thus created get attached with the previous block using the hash value generated for the previous block and a chain of blocks are created.

Note that the genesis block does not have any pervious hash value since it the first block in the chain. Bitcoin (Nakamoto, 2008) is an example of blockchain technology.

8.8.2.2 Directed Acyclic Graph (DAG)

DAG is another popular type of distributed ledger technology. A DAG is an alternative to the blockchain technology that works differently than blockchain but gives all the advantages that blockchain provides and some improvement on the limitations of blockchain. The DAG works on the zero-fee method, i.e., there are no fees required for the transaction in DAG network. A blockchain network might get slowed down with the increase in number of participants and will take up more time in confirming the transaction, but in DAG as the number of transactions increases, the time required in the transaction settlement decreases (AG, 2018). Unlike blockchain, which works like a linked list data structure, the DAG (directed acyclic graph) works like a graph data structure. It uses vertices and edges of the directed graph, which means the edges of the graph, will move in only one direction. The DAG is also an acyclic graph, which means there would be no cycle between the vertices (starting vertex will not be the end vertex).

8.8.2.2.1　How a DAG Process Works

Like the blockchain technology, DAG could also be used as a type of distributed ledger technology. In DAG, a transaction created gets stored on the nodes which represent a valid transaction on the ledger. Every time a new transaction is created it must validate at least previous two transactions. Once it is done the new transaction gets validated. Thus, DAG ledgers are advantageous in terms of size and scalability and are suitable for use cases where there is a high volume of transactions created per second.

A DAG ledger is free from blocks or chain; however, the ledger is still a peer-to-peer and decentralized network and uses the consensus algorithms for validating a transaction. Unlike blockchain, DAG functions without imposing any fees on the participants.

If we compare the two technologies, unlike blockchain, there are no blocks created in DAG, hence there are no miners in the DAG technology. There is no need of high-power usage as required in blockchain. DAG uses only a small portion of electricity than blockchain.

Nano (LeMahieu, 2014) was the first cryptocurrency created on DAG, originally known as Raiblock (Faraz Masood, 2018).

8.8.2.3　HashGraph

HashGraph is said to be the future of blockchain technology. The HashGraph technology, created by Leeman Baird in 2016, is patented technology and owned by Swirlds, Inc. HashGraph is a mathematical algorithm just like the blockchain. A HashGraph does not use the consensus algorithm like proof of work and thus the technology is faster as compared to blockchain and it can process around 250,000+ transaction per second (Muens, 2018). The HashGraph technology uses a system called Asynchronous Byzantine Fault Tolerance, which helps select the transaction that happened first.

In the HashGraph there can be multiple transactions that can be stored on the ledger at the same time. The HashGraph technology works on the goal of finding a consensus to find which node has created a transaction and the order of the transaction as well.

8.8.2.3.1　How HashGraph Technology works

Since the overall goal of the HashGraph technology is to find the order of transaction, consider an example where there are three participants in a network and all the three participants want to submit a transaction. Here comes the problem of selecting which participant submits the transaction first.

The HashGraph technology uses an algorithm known as gossip protocol (Alan Demers, 1987). A gossip protocol works the same way a broadcast happens in the network. A gossip protocol spreads the information in the network. A node possessing the information shares the information with all other nodes present in the network. The process continues until all the nodes get the information and thus synchronize the whole network with this information.

The gossip protocol works well and is a good way of sending the information to the network. Since the information is being synced with all the nodes in the network, the single point failure reduces in the network as all the nodes are sharing the complete same information. Thus, the HashGraph ledger is a decentralized ledger since all the nodes can directly communicate with the other nodes present in the network.

However, using the gossip protocol alone in HashGraph will not make the HashGraph technology a tamperproof technology. HashGraph should be in distributed ledger technology any malicious node can tamper or modify the information which may change the order of the transaction. Some advancement was needed.

Swirlds Inc. the inventors of HashGraph gave the solution of gossip about gossip protocol. Consider an example in which there are multiple participants. The first participant creates an event and gossips it to other participants telling that an event is created by him. He adds a timestamp to the event, and now another participant adds his timestamp to the event as metadata, which says the correct order of the transactions. Now in gossip about gossip protocol instead of one hash, two hashes are created where one hash gives the information most recent event of first participant and second hash gives the information about the recent event of second participant, and so on.

Thus, whenever there is a gossip from a network, it will include two types of information—one new information, and other information about recent gossips done in the network. Hence the protocol called gossip about gossip.

8.8.2.4   HoloChain

HoloChain is a framework building fully distributed peer-to-peer applications. A HoloChain is not a blockchain and is said to have a very advanced level of ledgers. A HoloChain takes an agent-centric approach rather than a data-centric as in case of blockchain (Harris-Braun, 2018). With the blockchain every node of the network maintains the same state of the network, whereas a HoloChain network does not use any consensus protocol and provides each agent with its own kind of system. Thus, by going with the agent-centric structure in HoloChain, each agent is maintaining its own history and its own hash chain whether the agent is posting any new information or any transaction it will get stored on its own hash chain

Like in blockchain network, all the nodes present on the network need to follow the global consensus and verify the complete network. HoloChain works in a bit different manner.

The HoloChain technology work on the concept used in the architecture of a hologram. Like in a hologram, in order to get a 3D image some specific light beams have to be there and collectively the light beams create the image. The HoloChain works the same manner and uses individual agent's ledger for creating the whole ledger.

In HoloChain every node present on the network keeps their own ledger and communicates with this ledger to other nodes through a unique signature.

8.8.2.4.1   How HoloChain Technology works

In HoloChain, each node will have their own ledger and this ledger is bind by some set of values called as DNA. This value of DNA must certify that any node present on the network trying to add any information on the ledger will gets validated.

A node sends some information to other nodes in order to get validated on the network. The nodes present will verify the node with the DNA, and if the DNA matches, the information can be sent to other nodes in the network.

However, if an unauthorized person tries to hack the network or tries to put false information on the network, the DNA would be different than the DNA present in the network for authenticated nodes. Once the DNA does not match or is dissimilar to the authentic one, the nodes will reject it and inform other nodes on the network regarding this unauthorized node.

8.8.2.5   Tempo (Radix)

A Radix is highly scalable, efficient, secure distributed ledger technology. The blockchain technology has some limitations, one would be scalability, by which the blockchain technology does not lend us well to mass market use. Second would be a bottleneck created due to the increase in transaction, which slows down the blockchain.

The developers of Radix named their consensus algorithm and ledger architecture Tempo. In Tempo computers are connected in a peer-to-peer fashion in order to communicate with each other. Like other DLT platform, Tempo also preserves the order of the information stored on the ledger. Tempo also gives the functionality of a timestamp and various other features as well. Unlike the blockchain technology, Tempo does not require any large or expensive hardware and can even work on the smart devices like mobile phone.

Every instance on Tempo is known as a Universe, and every event within the Universe is known as an atom.

8.8.2.5.1   How Tempo DLT works

The working of Tempo is slightly different from other distributed ledger technologies. In Tempo, any node on the network is allowed to carry a subset of whole ledger. This subset is known as shards. Any node carrying this shard/subset will get a unique ID for the subset of the ledger. Hence, there is no need for nodes to carry the complete ledger with them. This will make the network carry large amounts of transaction data and increase the scalability that was a limitation of the blockchain technology.

Tempo uses a logical clock for validating a transaction. In Tempo the node will record all the sequence of the event happening. So if two transactions arrived, it would be the duty of nodes to check which transaction arrived earlier and record that transaction on the basis of a logical clock.

8.9   PLATFORM FOR DLTs

There is multiple platforms available for different distributed ledger technology, which can be very useful in different use cases in multiple sectors of any organization. These platforms provide any organization an environment for using different technologies as per their requirement. Few of these platforms are discussed in the following sections.

8.9.1 HYPERLEDGER

HyperLedger started in the year 2015 by a limited number of developers who came from various sectors like banking, finance, supply, and data management. All these people have one common goal in mind—to make blockchain as a technology more accessible to the world. So, keeping these things in mind the developers have started testing between applications and secure blockchain networks.

According to Brian Behlendrof, executive director of HyperLedger, "HyperLedger is an open source community of communities to benefit an ecosystem of HyperLedger based solution providers and users focused on blockchain related use cases that will work across a variety of industrial sectors." HyperLedger can be thought of as a software that every individual can use to create a personal blockchain. Many different companies and organizations started using HyperLedger for their business in order to improve the business and the quality of the process.

The HyperLedger platform was created under Linux foundation and has grown rapidly in the area of blockchain. A user can either install the HyperLedger platform in a Linux machine or if using Windows, the user can create a virtual machine with Linux operating system.

HyperLedger fabric is one of the known and most-used platforms that is being used for blockchain technology. HyperLedger is an open source platform,which means anyone can use and download the platform, and that too is free of cost. HyperLedger provides a good level of flexibility, scalability, confidentiality, and resiliency.

A HyperLedger fabric is a permissioned distributed ledger, and not everyone can join the network on its own. When a user going for the installation of the HyperLedger platform, it provides all the prerequisite commands necessary for the working with the HyperLedger platform.

On HyperLedger, the participants directly get associated with the deal and it get updated on the ledger and notified. This maintains a good level of privacy and confidentiality.

A modular and extensible framework is developed under HyperLedger where the common building blocks can be reused. Because of this modular approach, developers would be able to experiment with different types of components whenever they evolve, and the individual component can be changed without altering the rest of the system.

When we talk about blockchain on HyperLedger, we also need to think about the transactions and how the transactions need to be complete especially when the transaction is finance based, then smart contract comes into picture. In HyperLedger, a smart contract code can be written in different languages such as Java, Go, and NodeJS.

There are different types of HyperLedger framework present, which could be used for different purposes.

- **Hyperledger** Fabric
- **Hyperledger** Iroha
- **Hyperledger** Sawtooth
- **Hyperledger** Indy
- **Hyperledger** Grid
- **Hyperledger** Burrow

- **Hyperledger** Caliper
- **Hyperledger** Cello

HyperLedger raises and promotes various business based on blockchain technology, which includes distributed ledger framework, smart contracts, client libraries, utility libraries, and other sample applications (https://hyperledger.github.io/, n.d.).

The following command needs to be executed while installing HyperLedger for the first time.

```
sudo apt-get install curl
sudo apt-get install golang-go
export GOPATH=$HOME/go
export PATH=$PATH: $GOPATH/bin
sudo apt-get install nodejs
sudo apt-get install npm
sudo apt-get install python
sudo apt-get install docker
curl -fsSL https://download.docker.com/linux/ubuntu/gpg | sudo apt-key add -
sudo add-apt-repository "deb [arch=amd64] https://download.docker.com/
    linux/ubuntu $(lsb_release -cs) stable"
sudo apt-get update
apt-cache policy docker-ce
sudo apt-get install -y docker-ce
sudo apt-get install docker-compose
sudo apt-get upgrade
```

8.9.2   R3 Corda

A lot has been changed from 1960s computers runs everything and is being used in every small or large sectors be it a banking and financial market or processing a billion of transaction in a day. But something has not changed yet. Suppose someone creates a transaction, before the transaction reaches to the receiver, the transaction reaches to multiple users in the network for authentication. At the end of the process both the sender and the receiver sees the same update.

What if there is simpler way for everybody? There is distributed ledger technology (DLT). The DLT is about sharing the data to keep the sender and receiver in sync but to make sure that no one can change or modify that data. There are a lot of platforms available today that uses blockchain technology. The problem with blockchain is that, everyone receives a copy of data or transaction which results in lots of duplicate data, which is hard to scale to meet the world's need. Privacy is another factor which is to keep in mind while using any platform. However, using encryption techniques, data can be encrypted but usually the persons sharing the data are only the sender and the receiver.

R3 was an enterprise blockchain company started in 2014. R3 made its first blockchain open source platform for business in November 2016 (Kelly, 2016) and named the platform Corda.

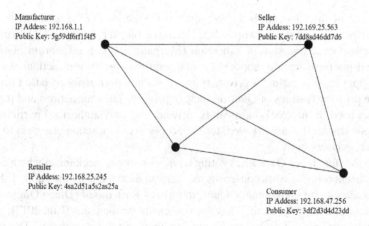

Manufacturer
IP Addess: 192.168.1.1
Public Key: 5g59df6rf1f4f5

Seller
IP Addess: 192.169.25.563
Public Key: 7dd8sd46dd7d6

Retailer
IP Addess: 192.168.25.245
Public Key: 4sa2d51a5s2as25a

Consumer
IP Address: 192.168.47.256
Public Key: 3df2d3d4d23dd

FIGURE 8.8   KYC process of Corda network.

Corda is a platform for permissioned network and only authorized nodes are allowed to join the network to communicate with each other on a need-to-know basis and update the shared information.

To meet the needs of business Corda network architecture it differently. Before joining the network, each node goes under a KYC (Know Your Customer) process in order to obtain an identity certificate and when they join the network, they publish their legal identity as their name, IP address, the public key information for network map service. The nodes then can use this network map service to transact with well-known parties in the network using private point-to-point messaging (Figure 8.8).

When there is any deal on Corda, data is put into a block with everybody else's. Each deal is individual. Like HyperLedger, on Corda smart contract can be used for trading just anything a person can think of. The information is sent to only those who need to know the information. Sometimes there is only a sender and receiver in the transaction process, and sometime you might want to plug in a service in order to check if the deal is fine or not, and sometime a regulator needs to be kept in a loop. With Corda, you are in control.

In Corda the famous double spending problem can be prevented by using notary pools (https://docs.corda.net/head/key-concepts-transactions.html, 2018). A notary pool is a set of nodes generally set of mutually distrusting node operating a Byzantine fault tolerant consensus algorithm. They will only sign a transaction if it does not represent a double spend attempt. Every transaction requires a signature for a notary to be valid. The nodes in the notary pool cannot see the contents of ledger update; they only see the hash of each transaction updating the ledger plus the index of the fact being constitute in the outputs of that transaction.

Corda platform has been made to integrate common tools and languages like Python, Java, and JDBC.

8.9.3 QUORUM PLATFORM

Quorum is an open source platform. It is a soft fork, very well-known public Ethereum blockchain developed by J.P. Morgan, an American investment bank and financial service company. Quorum is a good solution for the issues like privacy, scalability and performance of the network when Ethereum blockchain is used. Quorum is an enterprise-focused version of Ethereum (Morgan, 2019). The Quorum platform is an ideal platform for any application that requires speedy transactions with high throughput for processing of private transactions in a permissioned-based network.

The primary features of Quorum include privacy. The transactions and the smart contract used in the blockchain can be private, and only authorized participants in a specific transaction are allowed, for the access to a transaction or access to bodies such a regulatory body.

Another feature of Quorum is voting based consensus mechanism which do away from current proof of work consensus mechanism used today by the public Ethereum blockchain among many others. Quorum utilizes Raft-Based (Diego Ongaro, 2014) and Istanbul Byzantine Fault Tolerance consensus mechanisms (Lin, 2017).

In Quorum, peer or node gets the permission using smart contracts. This ensures that only authorized participant joins the network.

Another important feature of Quorum is the increased scalability and the network performance. This is to note that this very important feature of a Quorum platform has, especially when it is for the relatively slow performing and poor scalability public Ethereum blockchain.

Now let us see how it works and concentrate of the basic Quorum blockchain architecture of which consists of two main components.

Quorum node: This is the soft fork of what we know as the public Ethereum blockchain, making the Ethereum as base code and adding a thin layer on the top, which allows the modifications like using voting-based consensus mechanism instead of proof of work and allowing transactions and smart contract to be privately executed.

Constellation: This is a two-part system, in which the primary focus is to implement the privacy features of Quorum.

The first part of constellation is a transaction manager who is responsible for the privacy of the transaction. The duty of transaction manager is to store and allows access to the encrypted transaction data; exchange the encrypted pay load with other participants transaction managers, but the transaction manager does not have access to any sensitive private key.

The second part of the constellation is the enclave. The enclave works with the transaction managers to strengthen the privacy by managing the encryption and decryption in an isolated manner. The enclave stores the private key and essentially a virtually HSM (hardware security module) which is an encryption method.

8.9.4 Stellar

Stellar is globally open source protocols for payments. Stellar is maintained by stellar development foundation in San Francisco in the year 2014. Stellar has some key features that need to be discussed. A stellar platform is a multicurrency network, any currency, asset, or a token can be issued directly inside of a stellar network.

The transactions happening in the stellar network get confirmed in less than five seconds. The fee required for the transaction is nominal as only a cent will be paid for 10,000 transactions, for example, a sequence of transaction would cost around 150 million dollars in a traditional network, whereas in a stellar network the cost would be only 20 cents.

The stellar platform is much more scalable than other DLT platforms because the stellar can process thousands of transactions in a second. In the stellar platform, the participant has the choice of choosing the participant they trust on the network out of several members present in the network.

A stellar reaches the consensus using two types of consensus. One is called local consensus and another is called global consensus. Further local consensus is called quorum slice and the global consensus is called quorum. It can be said that when there is enough of local consensus it would be a global consensus or in other words, a quorum arises when there is enough of quorum slice in the network.

In a stellar consensus mechanism, it is not necessary in global consensus that every participant agrees with everyone in the network but simply agreeing with the neighbor is enough for everyone.

The cryptocurrency of stellar network is Lumens (XLM) (Brett Roberts, 2018). The XLM are the native asset of the stellar network. The XLM is responsible for the real-time transfer of the value in the network and it acts as a bridge currency between the digital-fiat assets that are issued by the anchors.

Stellar decentralized ledger can be thought of as a database; a database that can store more than just account balances and payments. Another use case for this database is storing the offers to buy and sell the assets. All these offers represent on a global order book called decentralized exchange (DEX) (McCollom, 2018).

The best way to understand the stellar DEX is to compare it with the other exchanges like centralized exchange and exchange build on Ethereum.

In a centralized the private key is stored on a centralized server, which we know had been hacked in past.

In Ethereum DEX the private key is stored in the smart contracts, but this also has some hacking history but in stellar, the user is not controlling their private keys rather the user is controlling their money.

A stellar has an order book and offer match making built in at the protocol level. The on-ledger offers are a simple addition to the decentralized database.

Thus, we have seen there are some difference between the stellar and the Ethereum platform. We can summarize the difference including the Bitcoin in Table 8.1.

8.9.5 MULTICHAIN

A multichain technology is another kind of open platform that helps the users in developing a private blockchain that could be useful for any organizations in their financial transactions. A multichain technology provides an API and a command-line interface, which helps in preserving and setting up of multichain (Sharma, 2017b).

TABLE 8.1

Comparison of Execution of Transaction on Platforms Ethereum and Stellar with Bitcoin

	Bitcoin's Blockchain	Ethereum	Stellar
Average transaction confirmation time	1 hour	15 minutes	3–5 seconds
Average transaction fee	$0.30 per transaction	$0.13 per transaction	20 cents for 1000 transactions
Transaction per second	3 transactions per second	7 transactions per second	3000+ transactions per second
Consensus mechanism	Proof of work	Proof of work	Stellar consensus protocol (SCP)

Objective of Multichain: In the blockchain environment the visibility of the blockchain should always be shared with the active users so that there would be no confusion among the participants which ensures the complete stability and control over the transaction. The proof of work consensus algorithm can be useful in the process mining. In a multichain, two or more parties can create their own blockchain but there would only single blockchain for entire bit coin network. The transactions happening in the multichain network cannot be accessible by any third party unless there is permission from the administrator. Whereas, the transactions happening with bitcoin are open to anyone but the proof of work consensus algorithm is optional in a multichain environment. The value of the Bitcoin is only because of the proof of work involved in it. Since there is no proof of work associated in multichain, the transaction cost is very low. Since the multichain network is a private type of network, there has to be valid permissions for the miners to be able to mine multichain blocks.

Hand-Shaking Process in Multichain: Hand-shaking is a process in which the nodes in the network connect and communicate with each other. The same process happens in a multichain environment when the nodes in the blockchain connect with each other. The nodes present in the blockchain identify each other with the IP address and the list of permissions a node have. With this each node can send message to other nodes; however, the P2P connection aborts if the process is not delivering any satisfying results.

8.10   APPLICATIONS OF DLT

8.10.1   TRADING

Trading is one of the most critical and crucial distributed ledger use cases. The DLT is preferred as a technology for trading purpose as in trading the financial transactions can be done using the cryptocurrencies. Trading is a kind of business that involves risk and sometimes emotional decisions too.

Sometimes some bad players of the trading business disrupt the market only for the purpose benefiting for personal gains. Distributed ledger thus can be very beneficial and can provide all the transactions in a transparent manner, eliminating lots of paper work and also reducing the reliability on the banking system.

Once the information is written on to a ledger it becomes almost impossible to manipulate the content in the ledger and preserve the assets in a more secured manner. A DLT also provides secure wallet where a user can store all of his digital assets without any risk.

8.10.2 MANUFACTURING

Distributed ledger technology can be useful in the production industry. A network that connects all the workers in the same network does ensure great outputs in a short span of time. The manufacturing industry found a solution in distributed ledger technology that might improve the efficiency of the system and reduce the cost.

With DLT, all the parties involved in the production industry come in to same network and all the transaction made in the network gets updated with all the parties. Thus, a distributed ledger technology can provide an ecosystem in the production industry that can look at the employees and make rational output-based choices.

8.10.3 SUPPLY CHAIN

Supply chain has always been an important part of any production industry. It is the supply chain that makes a product from a raw material. Many different organizations are now investing largely on the DLT, which might improve the quality of the product and the customer satisfaction.

With the distributed ledger technology many industries would be able to administrate the ongoing process from delivery of raw material to manufacturing and till shipping of the products.

Using DLT in supply chain industry customer would also be able to track the history of the product such as who was the raw material seller, where was the production, and when the shipment will be delivered. Using a ledger network manufacturer and the raw material seller would also get to know better the requirements of the customer.

8.10.4 HEALTHCARE

Healthcare is one more sector that is growing day by day. A patient's healthcare information is increasing, too. A technology is required that can store all the information of the patient, and the patient can retrieve the information regarding his/her health any time he/she wants to. The health information includes lab diagnostic reports, X-ray reports, MRI reports, CT scan reports, and many more healthcare reports. A patient could not carry these number of records to every doctor and hospital, so the patient needs to store all the health-related information in a single place.

Distributed ledger is the technology that provides a platform in which all the information can be store in a single ledger act as a database for the user. The ledger is digitally protected and only the patient can use the ledger for accessing related healthcare information. However, patient can allow some other user to access the information like a doctor or a hospital staff.

8.10.5 Government Services

Distributed ledger technology could act as a next generation technology in the government service like election. It would become necessary that the election process go in a transparent manner. But when the authority is under some person the things get tampered sometimes, some illegal activities may come in. All these activities would get stopped if distributed ledger technology is used.

With DLT, every vote gets stored on the ledger network and is difficult then to tamper with the vote. If the vote is tampered with, the authorized person gets the notification.

Every election process will be transparent to all the users connected in the network and the voting process will be fair and thus, the DLT will prove to be a trusted technology for the citizens voted.

Other government services like the law department, housing authorities, and so on can also come under DLT for improving the services and eliminating illegal activities.

8.11 CONCLUSION AND FUTURE SCOPE

With the introduction of DLT in today's world a new technology has arrived which has changed the way of communication and storage of information. The DLT has given the platform by which not only static but dynamic information can also be stored. The distributed ledger technology has showed us the direction that is different than a typical database and can be utilized in day-to-day activities. The benefit of DLT is more utilizing the technology for improving the growth, than just storing the information.

Seeing the benefits of DLT, many companies are already looking to shift their database on the new ledger technology and some others are planning to shift. The DLT has given the industries and other organization a platform by which they can improve in terms of their business and most importantly customer satisfaction. The industry or organization can choose between type of platform and type of DLT for the benefit of their industry.

If the DLT is keep on flourishing the way it is, it is inevitable that the technology would probably increase much more in the near future. Making everything saved in DLT would surely help both the owner and the consumer. The consumer will feel more secure, can trust on the seller and financial fraud, and data theft will become the story of past.

But for this the distributed ledger technology has to be much more efficient, and hopefully, in the near future we can see DLT get invented with much more efficiency and much more capacity to gather the information in a sequential form. Data is changing every day, so the ledger should be able to support and store the constantly changing data.

REFERENCES

AG, A. B. (2018). www.advancedblockchain.com/docs/DAG_ADVANCED_BLOCKCHAIN_AG.pdf. Retrieved from /www.advancedblockchain.com.

Alan Demers, M. G. (1987). Epidemic algorithms for replicated database maintenance. *Proceedings of the Sixth Annual ACM Symposium on Principles of Distributed Computing* (pp. 1–12). New York: ACM.

Arshdeep Bahga, V. K. (2016). Blockchain platform for industrial Internet of Things. *Journal of Software Engineering and Applications* 9(10), 533.

Bayer, D., Haber, S., and Stornetta, W. S. (1993). Improving the efficiency and reliability of digital time-stamping. In *Sequences Ii* (pp. 329–334). Springer, New York, NY.

Brett Roberts. (2018). www.nasdaq.com/article/2-game-changing-newcryptocurrencies-with-serious-backing-cm907959. Retrieved from www.nasdaq.com.

Consulting, A. R. (2019). www.globenewswire.com/news-release/2019/02/24/1741170/0/en/Supply-Chain-Analytics-Market-Size-Worth-Around-10-7-billion-by2026-Acumen-Research-and-Consulting.html.Retrieved from www.globenewswire.com.

Dhingra, P., Gayathri, N., Kumar, S. R., Singanamalla, V., Ramesh, C., and Balamurugan, B. (2019). Internet of Things–based pharmaceutics data analysis. In *Emergence of Pharmaceutical Industry Growth with Industrial IoT Approach* (pp. 85–131). Academic Press.

Diego Ongaro, J. O. (2014). In search of an understandable consensus algorithm. *USENIX Annual Technical Conference* (pp. 305–319). Philadelphia, PA: USENIX Annual Technical Conference.

Faraz Masood, A. R. (2018). An overview of distributed ledger technology and its applications. *International Journal of Computer Sciences and Engineering*, 422–427.

Fawcett, S. E., Magnan, G. M., and McCarter, M. W. (2008). Benefits, barriers, and bridges to effective supplychain management. *Supply Chain Management: An International Journal*, 35–48.

Harris-Braun, E., Luck, N., and Brock, A. (2018). Holochain: Scalable agent-centric distributed computing. GitHub [Электронный ресурс]. https://github.com/holochain/holochain-proto/blob/whitepaper/holochain.pdf.

https://hyperledger.github.io/. (n.d.). Retrieved from https://hyperledger.github.io/.

https://www.ibef.org/industry/healthcare-india.aspx. (2019). Retrieved from www.ibef.org.

Jain, V., and Bhatnagar, N. (2012). Four C's in supply chain management: Research issues and challenges. *Third International Conference on Emerging Applications of Information Technology (EAIT)* (pp. 264–267). IEEE.

Kelly, J. (2016). https://uk.reuters.com/article/us-banks-blockchain-r3-exclusive-idUKKCN12K17E. Retrieved from https://uk.reuters.com.

Kumar, S. R., Gayathri, N., Muthuramalingam, S., Balamurugan, B., Ramesh, C., and Nallakaruppan, M. K. (2019). Medical big data mining and processing in e-Healthcare. In *Internet of Things in Biomedical Engineering* (pp. 323–339). Academic Press.

Lamport, L., and R. Shostak (1982). The Byzantine generals problem. *ACM Transactions on Programming Languages and Systems*, Vol. 4 (pp. 382–401).

LeMahieu, C. (2014). Nano: A feeless distributed cryptocurrency network. pp. 1–8.

Lin, Z.-C. (2017). https://medium.com/getamis/istanbul-bft-ibft-c2758b7fe6ff. Retrieved from https://medium.com.

Madeira, A. (2016). www.cryptocompare.com/coins/guides/what-is-the-block-size-limit/. Retrieved from www.cryptocompare.com.

Margaret Rouse, M. H. (n.d.). https://whatis.techtarget.com/definition/consensus-algorithm.

Marr, B. (2017). www.forbes.com/sites/bernardmarr/2017/08/10/practical-examples-of-how-blockchains-are-used-in-banking-and-the-financial-services-sector/#475ad4771a11. Retrieved from www.forbes.com.

Martinovic, I. (2017). Blockchains: Design principles, applications, and case studies. European Social Fund and the Estonian Government.

McCollom, K. (2018). www.lumenauts.com/explainers/stellar-decentralized-exchange. Retrieved from www.lumenauts.com.

Mettler, M. (2016). Blockchain technology in healthcare. *18th International Conference on e-Health Networking, Applications and Services*. Healthcom, IEEE.

Morgan, J. (2019). www.jpmorgan.co.jp/global/Quorum. Retrieved from www.jpmorgan.co.jp.

Muens, P. (2018). https://medium.com/@pmuens/hashgraph-b79f901add20. Retrieved from https://medium.com.

Nabil El Ioini, C. P. (2018). Trustworthy orchestration of container based edge computing using permissioned blockchain. *The Fifth International Conference on Internet of Things: Systems, Management and Security*. Valencia, Spain.

Nakamoto, S. (2008). *Bitcoin: A Peer-to-Peer Electronic Cash System*.

Nick, S. (1996). Smart contracts: Building blocks for digital markets. *EXTROPY: The Journal of Transhumanist Thought*, 18(16), 2.

Papermaster, M. (2018). www.networkcomputing.com/network-security/blockchain-and-its-implementation-challenges. Retrieved from www.networkcomputing.com.

Peters, G. W., and Panayi, E. (2016). Understanding modern banking ledgers through blockchain technologies: Future of transaction processing and smart contracts on the internet of money. In *Banking Beyond Banks and Money* (pp. 239–278). Springer, Cham.

R3Limited. (2018). https://docs.corda.net/head/key-concepts-transactions.html. (2018). Retrieved from https://docs.corda.net.

Ralf Steinmetz, K. W. (2005). *Peer-to-Peer Systems and Applications*. Springer. Vol. 3485.

Sharma, S. (2017a). Cryptography: An art of writing a secret code. *International Journal of Computer Science & Technology*, 8491(1).

Sharma, T. K. (2017b). www.blockchain-council.org/multichain/multichain-technology/. Retrieved from www.blockchain-council.org.

Stuart Haber, W. S. (1991). How to time-stamp a digital document. *Journal of Cryptology*, 3, 99–111.

Tejal Shah, S. J. (2018). Applications of blockchain. *International Journal of Scientific and Engineering Research (IJSER)*.

Thompson, C. (2018). www.blockchaindailynews.com/The-difference-between-a-PrivatePublic-Consortium-Blockchain_a24681.html.Retrieved from www.blockchaindailynews.com.

9 Blockchain and Big Data in the Healthcare Sector

Yogesh Sharma, B. Balamurugan, and Firoz Khan

CONTENTS

9.1   INTRODUCTION OF BLOCKCHAIN TECHNOLOGY

The blockchain technology came into existence in the financial sector with the arrival of the cryptocurrency Bitcoin in 2008. Since then the technology has emerged multiple folds, and the technology along with other concepts is getting into various sectors. A blockchain can be considered as a structured database like a linked list, with all the data nodes connected to each other with a pointer.

A blockchain technology will be very useful where all the nodes in a network want to share their work and information on a common platform. Bringing the multiple authoritative domain on this single platform will make them work in a trusted environment so that they can **cooperate, coordinate and collaborate** (Vedpal et al., 2012) with each other in a decision-making process in the network.

In the traditional way of sharing the documents, one person writes a document that might contain some data or information and sends it to the receiver. The receiver again writes some data on that and then sends it back to the sender. With this kind of sharing there is a lot of time consumed, and the sender and receiver are not able to write the document simultaneously. Also, if the number of users will be more in this kind of network, it might take a huge amount of time. Now if we use some Google document application for sharing the information, the time will cut down to very less but again the problem is with the single or central authority. First, the network is still not decentralized and suffers from a single point of failure. This could be a problem if the central authority goes down or the server crashes. Second, every user needs sufficient bandwidth for the simultaneous update of the document. So, these are some of the disadvantage of a centralized system and necessity of moving the system on a decentralized system.

A blockchain technology is a **distributed** and **decentralized** technology. Distributed means everyone in the network collectively executes the job. Blockchain is based on the concept of **distributed ledger technology (DLT)** and is known as one of the types of DLT. In the blockchain technology there is no centralized/single authority or a node that has complete control over the chain, thus called as a decentralized technology in which there are multiple points of coordination. A network created using a blockchain technology is a **tamper-evident** and **tamper-proof** network. A blockchain is a digitally implemented network as all the records in a blockchain network are secured cryptographically.

In the blockchain network every node in the network maintains a local copy of the complete information. A decentralized system must ensure the consistency in the local copies with the node present in the chain so that the copies at every node are identical with all other copies on other nodes. Thus, all the local copies are getting updated based on the global information in the network.

9.2 A BLOCKCHAIN PROCESS

Figure 9.1 shows a typical blockchain process, in this case a transaction made by a user. The transaction could be of any type; it could be a financial transaction, or it could be just sharing information to the other nodes of the network. It is to note that all the nodes connected in the network have a ledger with them, which is a decentralized ledger and is used to store every transaction happening in the network. This means whatever transaction is happening in the network it will be updated to all the node present. Any node that wants to create a transaction has to write the transaction in a block. A block is a kind of container that can take up around 500 transactions on an average, although the size of a block is just up to 1 MB (Madeira, 2016). Once a node finishes a transaction, this information is added in the block. This newly created block will get a new block number and a unique ID attached to the block. This unique ID is known as a *hash value*. A hash value is a value created by the hash function, which is to make a transaction secure. This newly created block is now going to into the blockchain network where all other nodes are present, whose duty is to verify and validate the newly created block. Now if the transaction made in the block is correct and authentic and also has no false information, the transaction got accepted by the nodes of the network and based on a consensus algorithm applied on the blockchain by the creator of the network. Once a transaction is validated by the nodes, the valid block gets added into the block. All the blocks that got validated are added one after the another and form a chain called a *blockchain*. All the blocks that are added in the chain have their own hash value, and the hash value of the previous block makes all the blocks connected with each other. The first block in the chain has its own hash value and does not have the hash value of previous block. This first block in the chain of blocks is called as *genesis block*.

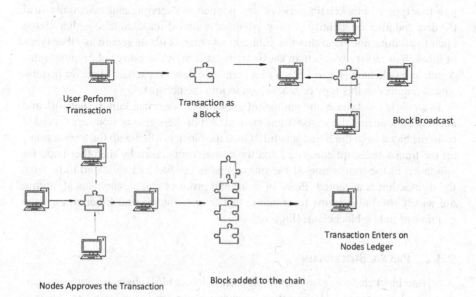

User Perform Transaction

Transaction as a Block

Block Broadcast

Nodes Approves the Transaction

Block added to the chain

Transaction Enters on Nodes Ledger

FIGURE 9.1 Blockchain process.

A blockchain works like a public ledger, which means anyone can join this type of blockchain. When a transaction is created by some node during that time it should ensure that this particular transaction, if it is a valid transaction, should get committed to the existing public ledger or blockchain; otherwise, that entry will not be there in the blockchain. Once the transaction is verified and validated by the other nodes of the network all the information should be consistent and get updated on all the local copies. The blockchain has to be secure enough that once the information is updated on all the nodes, no one could modify the data present in the nodes, or if some node is modified in the data and broadcast the information in the network, other nodes should be capable of detecting that there is some change in the data and the information must not be included. The privacy and authenticity are other important factors of a blockchain that need to be ensured as the data belongs to different clients.

9.3   TYPES OF BLOCKCHAINS

There are majorly three types of blockchain chain, public blockchain, private blockchain and a consortium blockchain. Depending on the architecture and the working of blockchain, different types of blockchains are used in various organization and industries. There are advantages and limitations to these types of blockchains but whichever type of blockchain benefits and improve the quality of an organization, it will be used despite of any limitation.

9.3.1   PUBLIC BLOCKCHAIN

The public blockchains are very well known due to the invention of cryptocurrencies like Bitcoin and Ethereum. Since it is a public blockchain, it means anybody can join this type of blockchain network and is open to everyone and essentially read the data and also the identity of the participant is almost impossible to gather. Being open to all does not mean that the public blockchain is not as secure as other types of blockchain. Every blockchain is essentially secure by the power of cryptography. A public blockchain is also known as a permissionless blockchain since no permission is required in this type of blockchain to join the network.

In a public blockchain any number of anonymous users can join the network and keep on announcing the transactions created. All of these transactions are recorded on to the block until the block gets full. Once the block is filled with the transactions, all the transactions are compared and the consensus is taken by the other node for validation of the transaction. If the transaction is verified and validated to be true, the transaction is accepted. Proof of work and proof of stake consensus algorithm are widely used algorithms for a public blockchain. Bitcoin and Ethereum are the example of public blockchain (Figure 9.2).

9.3.2   PRIVATE BLOCKCHAIN

A private blockchain is also known as permissioned blockchain because any participant who wants to join the network needs to get permission from the owner of the blockchain; all the permissions of the private blockchain are kept centralized.

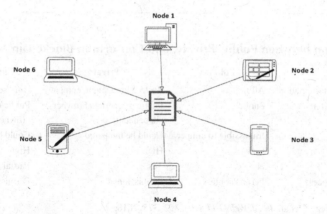

FIGURE 9.2   Public blockchain.

However, some people say that if blockchain is decentralized technology, then how come a private blockchain is a centralized one. Well, the organizations and many other sectors like banking and finance supports the private blockchain for being centralized. So, in a private blockchain the control is with the owner of the blockchain, and all the rule and regulations are made by the owner. It is the owner who decide who can join and who cannot. Since all the participants in a private blockchain are authorized participants so once the transactions are added in the ledger there no requirement of further audit. The details of all the transactions are visible only to participants of the network. When in a private blockchain the entire history of the assets can be tracked down and managed accordingly. All the participants have a ledger that gets updated every time there is any new information or transactions added in the block. The private blockchains are highly secure, and the cryptographically linked ledgers are almost impossible to tamper with (Figure 9.3).

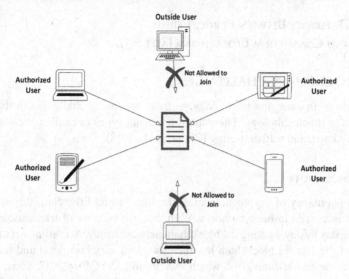

FIGURE 9.3   Private blockchain.

TABLE 9.1

Comparison between Public, Private, and Consortium Blockchain

Property	Public	Private	Consortium
Consensus determination	All miners	Restricted to single company	Only selected nodes
Read permission	Public	Public or restricted (owner dependent)	Public or restricted (owner dependent)
Immutability	Impossible to tamper	Could be tampered	Could be tampered
Efficiency	Low	High	High
Centralized	No	Yes	Partial
Consensus process	Permissionless	Permissioned	Permissioned

Source: Zheng, Z. et al., *Int. J. Web Grid Serv.*, 352–375, 2018.

9.3.3 CONSORTIUM BLOCKCHAIN

This type of blockchain is not a complete public blockchain, which means that it is not possible to join the blockchain like anybody. A consortium blockchain is actually controlled by pre-selected nodes in the network. A consortium blockchain is not considered as fully decentralized as everybody cannot just become involved in the network. However, they can be called a partial decentralized type of network. In a consortium blockchain the consensus mechanism is controlled by few selected participants, and it's the decision of these nodes whether to allow other nodes to read the records or not, although with this decision there might be a possibility that some wicked participant could tamper the records. The efficiency of consortium blockchain is high as compared to a public blockchain, and the consensus process in this type of blockchain is a permissioned one.

9.3.4 DIFFERENCE BETWEEN PUBLIC, PRIVATE, AND CONSORTIUM BLOCKCHAIN (TABLE 9.1)

9.4 BLOCKCHAIN CHALLENGES

Blockchain is an emerging technology, so there are lots of challenges in the implementation of this technology. There are three major types of challenges: scalability, privacy leakage, and selfish mining (Zheng et al., 2018).

9.4.1 SCALABILITY

With the popularity of cryptocurrencies like Bitcoin and Ethereum, the number of miners is increased in huge numbers. With that, the number of transactions is also increasing day by day, making the blockchain network heavy. According to (Liu, 2019) the size of the Bitcoin blockchain has increased to a very high level and reached to 226 GB at the end of June 2019, which was around 150 GB in 2017. Thus, the load a blockchain is bearing can be seen. All the transactions that are produced have to be stored on the blockchain for the purpose of validation of each transaction, and also if we talk about Bitcoin blockchain it can only process around 5–7 transactions per second. With these numbers of transactions, a blockchain cannot achieve the processing of a large number of transactions in real time. The size of the block is around 1 MB (Madeira, 2016). This size of the block is considered as very small when compared to the number of transactions that might cause delay for many transactions, but if the size of the block is increased it will again slow down the broadcast speed.

Thus, the problem of scalability continues to be the greatest challenge for a blockchain. However, in recent past some methods were proposed that might optimize the storage issues of the blockchain.

9.4.2 PRIVACY LEAKAGE

Many organizations are moving on to blockchain technology because the technology is believed to be secure and provide privacy to the users of the blockchain because every time a transaction is created it is supported by the address of the user and not the original identity of the user. However, in the recent past it has been found that the blockchain does not guarantee the privacy of the transactions made (Ahmed Kosba, 2016). If it is a public blockchain, every user can read the transactions made by other users and all the transactions, balances, are completely visible to the other users. The problem of privacy leakage arises because the other users get to know the address of sender and receiver because their exchange is being done on the same address, so the real identity of the user can be found out.

There are several methods proposed like Mixing (Möser, 2013), Mixcoin (Joseph Bonneau, 2014), Coinjoin (Maxewell, 2013) and CoinShuffle (Ruffing et al., 2014). These methods are based on the mixing and shuffling of the addresses of the receiver in order to prevent theft.

9.4.3 SELFISH MINING

The blockchain is inevitable to attacks from some miners, known as selfish miners. In the blockchain technology it is believed that if the nodes have 51% of the computing power then they can reverse the blockchain and thus can reverse the transactions. However, there are some recent researches that say if the node has less than 51% of computing power, still it would be dangerous for the blockchain.

In selfish mining, the selfish miner does not broadcast their mined blocks, and they wait for the private branch to get public. A private branch would get public only when some of the requirements are fulfilled. Now before the private blockchain gets open for public non-selfish miners are waiting for the private blockchain and wasting their resources on other branches whereas the selfish miners keep on mining the private chain without any interference from other miners, which enables the selfish miners to generate more profit than others.

In 2015, Saki Billah gave an approach to stop the selfish miners and proposed a Freshness Preferred Method and include an unforgeable timestamp, which might penalize the miners who are beholding the blocks (Billah, 2015), but the approach was vulnerable to a forgeable timestamp.

Then Saki Billah gave the idea of maximum acceptable time (mat). According to this approach a new block must be generated and received by the network with in maximum acceptable time for receiving a new block interval. So, within this mat, an honest miner would receive or generate a new block, and if the miner fails to generate a new one then it would create a dummy block. Thus, with this ZeroBlock approach the selfish miners would not be able to amplify their gain more than the expected.

9.5   BIG DATA

9.5.1   EVOLUTION OF DATA

We have seen how the technology evolved in last decade or so. Earlier we use to have wired or landline phones, but the technology advanced and we moved on to mobile phone. While working a computer we use to store the data on CD/floppy disk that were capable of storing a few MBs. Now that data has increased in large volume, and we now have the cloud storage for that. Take for example the use of mobile phones. We cannot imagine how much data we are producing every second, every action of a user performs even if one single picture is sent through the phone data gets generated. Now this kind of data generated is not in the format that a relational database can store and also the volume of the data has increased exponentially.

There are some other platforms that are giving large amounts of data, such as the IoT (Internet of Things) and social media. IoT connects the physical device with internet and makes a device smarter like smart TV, smart AC, etc. IoT devices includes various sensors that collect the data and can perform actions accordingly. So, we can see the volume of data generating when we have lot many sensors connected. According to (Fingerman, 2019) the IoT will grow to about $520 billion by 2021, and the total number of IoT devices is projected to be 75 billion by 2025 (Statista Research Department, 2016).

Similarly, social media is another important factor in the evolution of Big Data. Everyone is using Facebook, Twitter, etc., as social media. All the pictures posted, all the information shared, the user profiles, everything is data in these social networks. These social media sites also show us that the data is generating in various formats, but the data generated and stored is not a structured data, and the volume is quite large.

9.6   WHAT IS BIG DATA

Big Data is a term that is used for data sets generated and collected both in structured and unstructured form in such a high volume that it becomes very difficult to manage and process this data using the traditional database tools and applications.

9.7   THE 5 Vs OF BIG DATA

Big data is very beneficial to any organization, as the organization would be able to gather, store and can modify the enormous amount of data when the data is arriving at right speed and at right time (Figure 9.4). The five V's are volume, variety, velocity, value, and veracity. These are examined in the following sections.

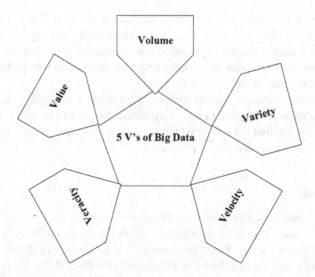

FIGURE 9.4 5 Vs of big data.

9.7.1 VOLUME

The volume of big data represents the amount of data generating. The volume of the data is increasing exponentially day by day. According to Forbes, 175 zettabytes of data will be generated by 2025 because the data is coming from different sources. The organization working in data collection has to deal with this huge amount of data. Sometime this large amount of data helps in improving the quality of the work of the organizations by predicting the results of the products.

9.7.2 VARIETY

Since there are different sources for the large volume of data, there can be different formats for these different sources. The different type of data can be broadly classified as unstructured, structured, and semi-structured.

An unstructured data could include data from different logs, such as an audio, video or an image file data.

In a structured format, the data is arranged in a structured tabular format with proper rows and columns and with the known schema of the table.

When we talk about a semi-structured type of data, all the data that is generated and received in various formats like Json, XML, CSV, TSV, or from the emails where schema is not defined properly are semi-structure data.

9.7.3 VELOCITY

Earlier when the users worked on a computer without internet access, they were working only on the stand-alone systems and not connected to the world. When the internet arrived, user data started generating, but the data was less in quantity and the speed of processing the data was very low. Then slowly and steadily as the number of users increased, the data generation also increased. It is when people started working on smartphones that the process of data generation reached a huge volume with more users sharing and posting of data.

There are few activities that are important for some users and need immediate action on the data, and thus the processing of the data has to be quick. So, for Big Data it is important that the rate at which the data is received, the processing of the data has to be on same rate.

9.7.4 VALUE

Now that the volume of data, variety of data, and velocity of data is done, we need to find the useful data out of the data collected. After the data is collected from various sources in different format, the useful data needs to be analyzed depending upon the type of data required for hypothesis or a particular process and that can help improve the growth of any business.

9.7.5 VERACITY

Veracity can be defined as the degree with which a user can trust the information for making a decision. Thus, it is important that whatever the value of data is extracted from the big data, it has to be accurate. As we can see in Table 9.2, there are lots of inconsistencies and some values are also missing. This happen because when a huge amount of data is put in, some data packets might get lost in the process. So, these missing values need to fill again, and the process of finding value starts all over again. Hence it is sometimes a challenge in finding the value of data as some business organization does not trust the value easily to reach to some decision.

TABLE 9.2
Sample Table Showing Inconsistency and Missing Values

Min	Max	Mean	SD
1.2	—	4.25	0.87
2.5	4.1	3.52	400,000
1,200	7.8	1.6	0.78
0.2	5.2	—	4.2

9.8 BIG DATA AS OPPORTUNITY

There are always some problems when dealing with the big data, but there are many advantages of big data as well and there are many opportunities that big data provide. There are multiple fields where big data can be used as boon and the small problems start getting solved as we start using big data in various sectors.

Before using the big data, we need to find out how the big data can be stored cost effectively. Earlier too much money was spent on storage until big data came into the picture; it was never thought of commodity hardware to store and manage the data, which is both reliable and feasible as compared to costly storage.

Today big data can be seen in every sector, organization, and industry be it a healthcare sector, a manufacturing, supply chain, retail or any other sector. Big data can help any industry to grow in many folds; however, it depends on how effectively the information has been extracted which could be advantageous for an organization. Edward Curry (2015) in his chapter discussed some of the potential use of big data in various sectors and discussed strategy that can increase competition among various industries.

9.8.1 BIG DATA IN FINANCE AND BANKING

If we talk about the opportunity of big data in finance, generally the financial industries hold a huge volume of data in unstructured form because whenever the data is generated and collected, only 10% of data is structured and remaining 90% of this data is unstructured. This kind of data is largely under-analyzed, and majority of the firms are not using this unstructured type of data. Most financial companies do not prefer to combine both the unstructured and structured data. With the traditional technologies like relational database management it is a challenge to process a huge volume of data growing every second, which would make the analysis of the data very difficult. The financial firms are focusing on different innovative solutions and strategies with which the risk assessment could be done and could be helpful in forecasting and trading. The financial services are interested in collecting the consumer's data, which can be used in predicting the consumer's behavior patterns. In retail banking, big data is playing an important role in order to tackle various challenges such as risk management, fraud detection/fraud avoidance and future price trends. With big data, now the banks could save more information related to a customer for a longer time, which makes it easy for the banks to track down the history of those high-risk customers who are in the process of making any kind of loan from the banks. The big data analytics could be used in order to find out what kind of facilities a customer is looking for; it has been seen that the quality of service is one of the most prominent factors customers prefer when switching to other banks. Big data could be helpful to banks in such a situation and provide the bank with the possible reasons why customers frequently switch to other banks.

9.8.2 BIG DATA IN EDUCATION

In the education sector, big data plays a significant role in finding the interest of the students. Marín-Marín et al. (2019) proposed an objective to analyze the scientific production on big data in education. The purpose of big data in the field of education could be to analyze and improve the teaching/learning process among the students. Big data as a tool could be useful in collection of large amount of student data for the analysis of interest of the students in different subjects and the method used. This could be beneficial for students, and it would increase the efficiency and quality of learning among them (Figure 9.5).

Drigas and Leliopoulos (2014) discussed the benefits of big data and open data in education. With the open data the parents and students would be able to search for good schools or good education-related programs. The student would be able to find jobs from the pool of jobs that was not possible earlier when only limited jobs could be shown to the candidates. SBooks and other materials are available online, and students can test their knowledge by taking exams that consist of a large number of questions. Online data available gives a student multiple options from reading different books, preparing for interviews and preparing for competitive exams. Big data helps an institution in identifying the progress of their institute in terms of achievements and status and also the weakness and method to improve as compared to other institutes (Bamiah, 2018). With big data as a tool, any institute can plan out the strategies necessary to improve the quality of education and other activities that will improve the performance of both teachers and students at the institute.

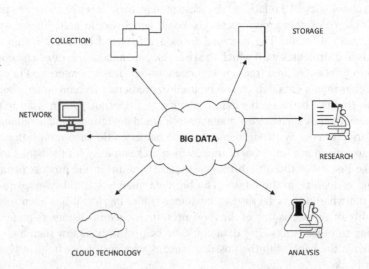

FIGURE 9.5   Big data in education.

9.8.3 Big Data in Farming

The role of big data could be beneficial in the agriculture sector in terms of both smart farming and related supply chain management (Wolfert et al., 2017). Big data technology with the IoT could produce a huge amount of data related to farming and agriculture. The data generated and collected with the smart sensors could be of very large volumes and can be useful in the prediction and decision making process that might be fruitful for any type of farming (Figure 9.6).

The big data technology has come up as a game changer for any business that would definitely improve the business process of any organization. Monitoring of crops could be done using big data technology; farmers can protect the crops from diseases, and proper pest control can be executed based on the history of the crop by feeding both previous and current data into a predicting system. The right period of the crop can also be checked to protect the crop from extreme weather conditions, which was not possible earlier without the big data technology. The farmers now have the choice to sell the crops to a good buyer from a list of buyers maintained and not limited to or dependent on a single buyer who buys the crops on his own conditions. The list of buyers could also show which buyer is going to buy which crop and prices could also be mentioned. The risk assessment and analysis using big data could be beneficial in farming, which could help a farmer take the appropriate actions using the real-time data so that there would be minimal damage to the crop.

Farmers can check the previous production with respect to demand and supply of the crop. This information can be used by the farmer for the production of the crop, which would improve the farming operations.

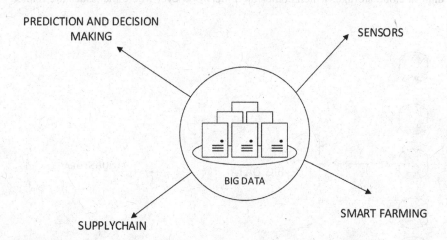

FIGURE 9.6 Big data in farming.

9.8.4   Big Data in Cloud Computing

Merging big data with cloud computing is a powerful combination and it can transform any organization (Chan, 2018). Big data, as the name says, is a huge volume of data; for storing this huge volume of data, there has to be some storage that can accommodate any quantity of data. The traditional storage system does not have enough capacity to store large volume of data and neither can they able to process this large volume of data, especially when the data is coming at fast pace. So there has to be a better method and place where the data is processed and analyzed. That's when a data storage method comes to the rescue; it's called cloud storage where any amount of data can be stored (Figure 9.7).

Big data can have the volume of petabytes and exabytes; it may soon reach to zettabytes of data, and since a cloud has the scalable environment that can store any volume of data, it is possible to deploy any kind of application that may boost up the analysis of any business by making the big data easy to access and streamline. There are various companies that provide cloud storage, and a good company ensures that the work of any business or organization always goes perfect without any problems. Earlier, many companies used to invest huge amount in setting a big data center for storing the data, but with the cloud computing the responsibility of data, storage is taken by the cloud providers, and the user company just needs to pay the service charges. A cloud storage is a dynamic storage that means the cloud storage could expand or reduce depending upon the volume of incoming data. Awodele (2016) discussed some of the issues and challenges associated with big data in cloud computing. One of the challenge big data and cloud storage come up with is the rate at which big data is being generated every second. The resources associated for the processing of this big data not able to match with the rate of flow of data because of limited bandwidth of network connections.

The issues that can come across with big data are security issues related with the data in cloud storage, which comes up at network level where the issues are related

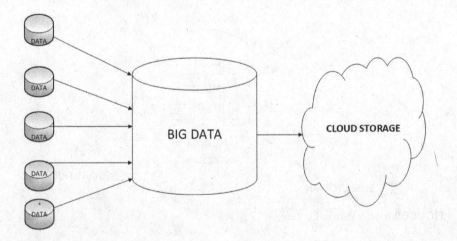

FIGURE 9.7   Big data in cloud computing.

to the network security. There are challenges at the user authentication level that includes the encryptions or decryption techniques and the authentications methods used for accessing the cloud storage. Protection of data is also a challenge in the cloud storage.

Big data analytics is another thing that is an important process in analyzing big data. Many websites use big data analysis in order to understand the user's requirements and thus helps to improve the quality and customer satisfaction.

There are some other benefits that big data analytics provides that can help a business grow and find better insight of the data.

Cost Reduction: Using the big data analytics, a cost-effective storage system can be maintained for huge data sets.

Faster and Better Decision Making: Using big data analytics, a company or an organization can gather quick and reliable information that can be very useful in making quick decisions for the betterment of the company.

9.8.5 BIG DATA CHALLENGES

There are quite a few challenges when dealing with big data. The challenges could be engineering challenges, and they could be semantic challenges. The engineering challenges could be related to storage or queries, whereas the semantic challenges could be dealing with the huge amount of unstructured data. There are a few more challenges mentioned by (Jones, 2012) regarding the management of big data.

• The processing of big data is challenging when dealing with a high volume of data with low-power processing tools.
• It could become difficult to analyze the data in real time.
• Sometimes it is difficult to design a scalable storage system where efficient data mining techniques could be applied.

Furthermore, there are some other challenges with big data (Jianqing Fan, 2014).

• Big data could be a complex data, when the data is received from multiple sources.
• Measurement errors, missing values, and outliners in the big data arise due to some noisy data.
• Dependent data could be big challenges where in some values are dependent on other values.

Mardis (2016) has discussed a few challenges related to big data. Big data for any company can either be generated by a company, or it can be collected from a third party. If the big data is being generated within the company, the company knows the processing and computational cycles of the data; but if the data is being collected from outside the company, then it would be a challenge for the company as the company needs to know the volume of the data being collected and also the computational cycle required in order to compute this data. Furthermore, the type of the data and the algorithm required also needs to be checked. The accessing and sharing of big data are also a big challenge. The data collected from different sources might be in different formats and from different platforms that might be challenging for accessing the data. The sharing of the data might involve some legal tasks or documents that need prior approval from the signing authority.

There could be some other challenges that come across while uploading the data. The rate of uploading the data should match with the rate at which the data is received.

There are some challenges associated with the security and privacy of the big data. A company needs to work upon the security and privacy of the big data either generated or collected. The security issues come up when the data is being accessed not only by the authorized users but when some unauthorized users also access the data of any company. In the same manner, the privacy of the data is also a concern from the malicious users when the data is of electronic medical record then the security and the privacy of the data is of major concern. The healthcare data can be misused by any person like it could be misused by an insurance company. Another issue of leaking of medical data and the modification or some tampering of medical record could also be done.

The storage and management of the big data is another challenge that comes across while dealing with the big data for any company.

9.9   INTRODUCTION TO HEALTHCARE

According to Winston S. Churchill, "Healthy citizens are the greatest assets any country can have."

Every country has their own healthcare system, and it became the responsibility of each country that they should make better plans to promote basic healthcare facilities for the people of the country. However, some countries have seen incredible improvement in the healthcare sector in some last decade or so. But still there are some gaps or flaws in the healthcare system.

According to the World Health Organization (WHO), health can be defined as "… the state of complete physical, mental, and social well-being and not merely the absence of disease or infirmity." There is no independent definition of health, and it depends on many factors. A healthcare system can be divided into a public healthcare and private healthcare. Public healthcare is generally run by government agencies and deals with either some groups of people or some individuals whereas private healthcare can be a profitable or non-profitable (NGO).

The current healthcare system in India is composed of primary healthcare, which deals with the primary healthcare centers and some subcenters. Healthcare can also be provided by the hospital or the health centers that include the community health centers; some rural and district hospitals may also come under the health center.

The healthcare sector has grown so much in some recent years and so has the data of the patients. The healthcare system is generating huge volumes of data every day as compared to any other sector. Earlier, the patient's data was maintained in records books t by the hospitals or some healthcare system. This information was not only difficult to maintain but also difficult when searching in previous data. With the introduction of computer and databases, the management of healthcare data becomes easy for any healthcare system.

The huge volume of data that is generated can be analyzed using big data analytics, which will be useful for many researchers and health practitioners to make the decisions regarding the treatment of the patient. The patients can now store their health records electronically, so that they can access their health records anytime and anywhere they are needed. Such records are called electronic health records (EHR) (Tom Seymour, 2012).

9.9.1 ELECTRONIC HEALTH RECORD (EHR)

A medical record is the collection of a person's healthcare information. These medical records are used to store the complete medical history of a person. The history could include the patients' healthcare, previous diseases, different trends of diseases, medication done for different diseases, types of allergies with any medicine, and so on. Thus, these healthcare documents were used to physically store file or medical reports.

In around 1960s, Lockheed developed an electronic system to store the medical history of a person; at that time it was called the Clinical Information System. Then, in the 1970s, the federal government started using the EHR and in 1980s the use of EHR increased and physicians and hospitals started using EHR (Marquez, 2017).

Electronic health records are the records of a patient that are stored electronically. With the electronic health record, a patient does not need to carry all of his or her records to different doctors and hospitals. The patient can access the records in real time. With the electronic health record, it would become easier to do the medical documentation, track the health record, and make the billing and coding process easier. An EHR is also known as electronic medical record (EMR), and both can be used interchangeably. There is another term called the personal health record (PHR). A PHR is a kind of record that can be changed or modified by the person (Roman, 2009). EHRs are protected under HIPAA (Health Insurance Portability and Accountability Act) and cannot be modified, whereas PHR are not covered under HIPAA.

An EHR can consist of the health records of the patient, but it can also contain any document that is important for the EHR, such as clinical information that may include the list of medication, list of allergies with any food item or any medicine, list of immunizations done, laboratory or pathology reports, surgery reports if any had done earlier, etc. An EHR could also consist of any administrative information that might include insurance information, appointment history, diagnostic and procedure applied, billing information, and some emergency contact information.

There are eight core functions of an EHR. In 2003, the U.S. Department of Health and Human Services (HHS) asked the Institute of Medicine to set standards for the EHR systems. In response to the Institute of Medicine gave the outline of the eight core functions of an EHR (Clancy, 2003).

- Health information and data management
- Results management
- Decision management
- Electronic communication and connectivity

- Patient support
- Administrative processes
- Reporting and population health

9.9.2 RISKS INVOLVED WITH EHR

With the advancement with EHR there might be some liability risks involved with the electronic health records (Minal Thakkar, 2006) where the privacy and inaccurate patient information is among the greatest of the risks involved in an EHR. Some risk includes template documentation making accurate correction to the document. If the user is not careful and finds that the record is edited completely, the user is left with mistakes and the problem then leads to multiple issues. The issues could be like any time when a person enters the information in the record over the prior information from an earlier visit or from some another portion of the record and tries to make that record reflective of the present visit. Any government agency might consider it as fraud; the government might think that they have been charged for the services that were actually not provided. Whenever the patient has the documentation in the record that is inaccurate, it might affect the credibility of the provider. But if it is absolutely necessary, only the update can be done, such as if there is an update in the diagnosis of some medical problem or if some new allergies come up with some food or medicine.

There are some other risks that are involved with the EHR. If the EHR system is not available due to some technical issues with the system (Minal Thakkar, 2006) or the online system, it could create a problem for any patient if the patient's doctor wants to track the history of treatment done a year ago. The security of the record is another major issue of concern because the EHR of any patient can be used for any purpose without the prior knowledge of the patient.

Now, since the patient health information is increasing day by day, there might be a problem of storage of the record, privacy of the record and also protecting the record from being modified or being used by some unauthorized person. Blockchain technology provides a platform where all problems that exist with the electronic health records can be avoided. Using blockchain technology in healthcare would be a good idea for the patient health record with the benefits of security, privacy and the interoperability of the patient's healthcare record (Mian Zhang, 2018).

9.9.3 CHALLENGES TO ELECTRONIC HEALTHCARE RECORDS

The EHR did benefit the patient to record their health data in a single place so that it can be accessible anytime, anywhere. With EHR, a patient can access the history of diagnosis, treatment and medication received from different doctors in different hospitals. EHR can also benefit the physicians as it may help them to track down the previous records of their patients and also helps in checking if diagnosis already done on a particular disease and will avoid the repetition of the tests.

But if there are some benefits of the EHR, there could also be some challenges associated with the electronic health records.

First, the challenge could arrive in the hospital where there could be a risk related with the privacy of the healthcare record; there are administrative staff who can steal the data and use the health record in some other manner. There are some hospitals or small healthcare institutes that find the use of EHR system to be costly to implement because they do not get the funds for the system (Chao et al., 2013).

Second, the patients storing their health-related information may come across privacy issues of their health record.

Third, physicians using the EHR are not comfortable using the system. Some of the physicians might not be comfortable working with the system, and some physicians find the system slow without the template and guidance necessary to use this type of system (Chao et al., 2013).

Interoperability is another big challenge for the EHR system because there are different EHR having different codes, technical specifications, keywords and different functionality. There has to be a standard format for every EHR system built in so that multiple systems are used interchangeably for the system work effectively (Reisman, 2017).

9.9.4 SECURING EHR USING BLOCKCHAIN

The electronic health records are the confidential and private information of any person. People don't discuss or want someone else looking into their health-related information. The EHR needs to be secure as there were many cases of data leak or security issues related with it. According to HHS around 8 million people were affected by a recent medical data leak (Merisalo, 2012): Thus, there has to be a system of a technology that can secure the medical data of the patient and not allow any unauthorized persons to access, read or modify the health record. Blockchain technology is one such technology that is being used in various other sectors to maintain the privacy and security of important information.

Jayneel Vora (2018) has developed a blockchain framework for better maintenance and storage of the healthcare data of the patient. The framework also is meant for the securing the health record and better accessibility for the person authorized to access the health record. The cipher manager and encryption technique are used for sending and receiving the health information over internet, which will reduce the unauthorized access to the health record by some outsider. In this framework each patient was given an Ethereum address and an identifier, which will make the accessibility difficult for an unauthorized person. The patient is given the full control and the access of the health record. The patient has the ownership of the record and only patient can allow the access of the healthcare data to other unauthorized users.

Healthwizz is a mobile phone app created by healthwizz.com (2019) where a patient can collect the health records from various sources like wearable devices, lab reports, X-ray reports, from a doctor or from a hospital and can sync their health records from other devices or application to the Healthwizz application. The health record thus put in the application is very secure and private to the patient, and the application does not share the health record with any unauthorized person. However, the application allows the patient to share the health record with other physicians, fitness trainers, or other users, as desired by the patient, on the secure blockchain platform. The patient can access their health information wherever they go and always own a recent copy of it, even without an internet connection. Since the blockchain cannot be hacked, if the patient wishes to share the health information with a physician, only that transaction gets locked on blockchain and the medical data shared on a secure channel.

According to Raj Sharma, CEO HealthWizz, today there are government mandates in place where consumer have the right to aggregate the medical records that are scattered all over the place, be it with doctor, hospital, insurance company and more data is generated day by day. The tools used should be able to handle this big data, plus the tools should be able to aggregate the data and make it available on users' mobile phones so that it could be accessible on blockchain platform. Using this process, the record can be shared securely either in case of an emergency or if the user is traveling.

With the health records on the blockchain, the records can be accessible from the user's mobile phone. In case the user is injured or unconscious, the biometric of the user can be used to access to the application and can view the health record. However, other user cannot transfer the health records as they are secure with the private key of the patient.

Rui Guo (2018) in this paper an *attribute-based signature scheme with multiple authorities* (MA-ABS) was used, which preserves the privacy and immutability of the electronic health records. In this scheme the patient approves the message based on the attributes and provides only that information which the patient wants to share or allow the access. Multiples authorities will generate and provide the public or the private keys of the patient, thus avoiding the problem of keys held by the third party. In the paper the author has demonstrated the attribute-based scheme is secure for the healthcare records in term of both the unforgeability and provides perfect privacy of the attribute-signer.

The EHR are scattered all over the place, from hospitals to doctors, which hampers the sharing of the data and increases the risk of being misused. Jingwei Liu (2018) has proposed a blockchain-based privacy-preserving data-sharing (BPDS) for EMRs. In this the patient's data is stored securely on a cloud so that the risk of data leak could be avoided, and indexes are reserved in a tamper-proof manner on the consortium blockchain, which ensures that the EHR should not be modified by unauthorized user. The data sharing, however, could be done by using the smart contracts that allow the access permission of the health record. The author also used CP-ABS-based access control mechanism, and the data extraction is based on the signature-based scheme that would help make the health record more secure for the patient and thus preserve the privacy.

Huihui Yang (2017) has proposed a blockchain and MedRec-based approach and implemented signcryption and attribute-based authentication that ensure the secure sharing of health records. With the implementation of the approach, the patient can view the complete health record that was scattered earlier in different places. By using MedRec the health record of the patient can be stored securely and kept safe against tampering, whereas with the sign encryption approach the authenticity of the health record of the patient can be verified and allows flexible data sharing.

9.9.5 DISTRIBUTED LEDGER FOR HEALTHCARE AUTHENTICATION

The electronic health record of a patient is stored in a healthcare application. These healthcare applications need not to be of same architecture (Angelica Lo Duca, n.d.). Different healthcare institutions such as hospitals, healthcare sector, doctors, fitness center, and so on, do not have the same application for the healthcare, and different institutions store healthcare records differently, sometimes not referring to those healthcare records. That means there is no use of extracting the history of the patient because it would be stored in different manner in other applications. Even if the electronic records of patient are stored in same institute, they would not be same until the time the record is not stored on a common platform or application. Thus, it would become very difficult for the patient to carry the history of records and diagnosis to every institute or healthcare provider. So, there has to be a platform where the records can be stored and accessible from anywhere.

Distributed ledger provides such a heterogeneous kind of platform that would be necessary for the patient to store the complete health records. A distributed ledger provides an interoperability that allows the health record to be stored on any platform, that too in a secure manner.

The patients are provided with public/private key to access their health record, which not only provides the security to the record, but it also gives the authority to the user to use the record whenever and wherever needed. Storing the health record on a distributed ledger also helps doctors track down the history of the patient regarding the medication and treatments done earlier.

With the use of distributed ledger any update done in the medical record of the patient—such as a medication, treatment or diagnosis update—could be done on the ledger and will update all the existing records present at the institute, hospital, doctor's office or for a fitness trainer.

9.10 DATA RETRIEVAL USING BLOCKCHAIN

There are some types of blockchain in which the user needs to analyze the blockchain in order to find the correlations in the data with the hashes and the transactions confirmed. So, there is a need to extract or retrieve the data from the blockchain. Ethereum and Bitcoin are two such blockchains from which the data could be extracted in order to review the content of the blockchain. There are some open source analytics tools by which one can extract the data from the already-deployed blockchain. For retrieving data from Ethereum blockchain the tools are QuickBlock, it is the collection of tools and used for retrieving the Ethereum data. EthSlurp is another tool that is useful in extracting the transactions for some specific address or a smart contract into a predefined format like CSV or a text file (Siddiqui, 2018).

A QuickBlock or QBlock is a collection of tools that consists of software libraries, tools, and applications that are useful in retrieving the data from an Ethereum blockchain. A QBlock speeds up the data retrieval and high level of information is extracted using QBlock. A QBlock is a maintenance free and fully decentralized software. When storing the data on an Ethereum blockchain a data should be optimized for retrieval before storing it on to the blockchain. Thus, a QBlock is useful in retrieving the data from the Ethereum blockchain and maintains fully decentralized feature.

Ethslurp is another tool like QBlock, and works for an Ethereum blockchain, which allows the user to extract the transactions for a specific Ethereum address that includes smart contracts. These transactions can be stored in a particular file format like .csv or .txt.

For example, *"ethslurp (some hash value)"* would return the complete transaction from an Ethereum blockchain.

Similarly, *"ethslurp -f (hash value)"* the ethslurp command can also be customize. This command allows a user to include any field in the transaction or any field to drop, the format of the data files and also the summary field to be included.

"ethslurp -f: txt (hash value)"
"ethslurp -f: csv (hash value)"
"ethslurp -f: html (hash value)"

These commands allow the user to extract the transaction information in particular format, if the user wants to store the information in. csv, .txt or in an html format.

9.11   DATA SHARING USING BLOCKCHAIN

The information stored on a blockchain network is always confidential data. A blockchain network provides security, authenticity and security for the data stored. However, it depends upon the type of blockchain, if the data stored is accessible to the other nodes of the blockchain or nodes that are not part of the blockchain network. If the blockchain is a public blockchain then any user can access the data and is open to any user. But, if the blockchain is a private type of blockchain then the access would get limited and depends upon the owner of the blockchain.

There are some use cases in which the records present on the blockchain demands sharing of records. Ajay Kumar Shrestha (2018) has suggested an approach with a blockchain-based model for collection of research data, providing the accessibility, maintaining complete and updated information, and verifiability of records with access of records, sharing of records and using the record. With this approach, the user will get the ownership of the record, with digital tokens, acknowledgement and sharing of record to only those who need the data.

In the proposed solution by Ajay Kumar Shrestha (2018) an Ethereum blockchain is being used for sharing the data with the participants already registered in the private blockchain. In the existing blockchain the concept of smart contract was used, which can get activated once a user gets registered on the blockchain with basic profile and public wallet address. The idea of using the smart contract is to get some incentives in exchange of sharing the data. The smart contracts are automated code that gets executed automatically when a condition gets satisfied. So, with the sharing of data through smart contract the owner will get some incentives.

Qi Xia et al. (2017) proposed a blockchain-based data sharing (BBDS) framework that talks about the challenges faced for the accessing and sharing of critical information that is stored in the cloud using the immutability and built-in properties of the blockchain. The system used a private blockchain, which allows only authorized users to access the content of the blockchain and the logs of the users gets created in the blockchain. The system created allows the user to request the data from the common ledger present in the blockchain. In the proposed solution a scalable and a lightweight blockchain is constructed which increased the efficiency of the network and allows data sharing, which is secured and protects the critical data.

Alevtina Dubovitskaya (2017) has proposed a framework for EMR data sharing and managing. The framework was implemented in order to ensure the security, privacy and accessibility of EMR data. The proposed framework focuses on four entities namely, doctor, patient, database and a node. Both the doctor and patient are registered with a membership service in order to authenticate the two entities. The membership service also includes the generation of key for signing and encryption for all the users. A functionality of symmetric key is used both for encryption and decryption of data. The symmetric key is used when a patient wants to share the health record with the doctor; an encryption key is also shared with the corresponding doctor. However, if the patient key is leaked or hacked by some other user, a patient would get the information regarding the same and could create a new key and again shared it with his doctor.

The health information of any patient could be of huge volume, for that a cloud storage could be used for storing such a huge volume, although if the record is of small size, it could be stored on a local database provided in the hospital.

The role of the node in the framework is to adapt the related chain code of the customer. All the nodes present will get updates regarding all the transactions happening in the chain and the leader of the nodes organizes all the transaction in the block and applies an appropriate consensus algorithm.

REFERENCES

Ahmed Kosba, A. M. (2016). Hawk: The blockchain model of cryptography and privacy-preserving smart contracts. *Symposium on Security and Privacy (SP)* (pp. 839–858). San Jose, CA: IEEE.

Ajay Kumar Shrestha, J. V. (2018). Blockchain-based research data sharing framework for incentivizing the data owners. *International Conference on Blockchain* (pp. 259–266). Springer.

Alevtina Dubovitskaya, Z. X. (2017). Secure and trustable electronic medical records sharing using blockchain. *AMIA Symposium*. AMIA Annual Symposium proceedings.

Angelica Lo Duca, C. B. (n.d.). How distributed ledgers can transform healthcare applications. Retrieved from www.iit.cnr.it/sites;www.iit.cnr.it/sites/default/files/LoDucaBacciuMarchetti-final.pdf

Awodele, O. (2016). Big data and cloud computing issues. *International Journal of Computer Applications*, 133(12), 14–19.

Bamiah, S. N. (2018). Big data technology in education: Advantages, implementations, and challenges. *Journal of Engineering Science and Technology*, 13, 229–241.

Billah, S. (2015). One weird trick to stop selfish miners: Fresh bitcoins, a solution for the honest miner.

Chan, M. (2018). Big data in the cloud: Why cloud computing is the answer to your big data initiatives. Retrieved from www.thorntech.com;www.thorntech.com/2018/09/big-data-in-the-cloud/

Chao, W. C., Hu, H., Ung, C. O. L., and Cai, Y. (2013). Benefits and challenges of electronic health record system on stakeholders: A qualitative study of outpatient physicians. *Journal of Medical Systems*.

Clancy, D. C. (2003). *Key Capabilities of an Electronic Health Record System*. The National Academies Press, Washington, DC.

Coughlin, T. (2018). 175 zettabytes by 2025. Retrieved from www.forbes.com;www.forbes.
    com/sites/tomcoughlin/2018/11/27/175-zettabytes-by-2025/#72badbd85459.
Drigas, A. S., and Leliopoulos, P. (2014). The use of big data in education. *International
    Journal of Computer Science*, 11(5), 58–63.
Edward Curry, J. M. (2015). *New Horizons for a Data-Driven Economy: A Roadmap for
    Usage and Exploitation of Big Data in Europe*. Springer, Berlin, Germany.
Fingerman, A. (2019). The IoT 2019: Many "things" ahead, but here are four predictions.
    Retrieved from www.clickz.com;www.clickz.com/whats-ahead-for-iot-many-things-2019/
    226144-2/226144/.
Health Wizz. My body, my data. Aggregate, organize and share medical records. (2019).
    Retrieved from www.healthwizz.com/#Message-2.
Huihui Yang, B. Y. (2017). A blockchain-based approach to the secure sharing of healthcare
    Data. NIK-2017.
Jayneel Vora, A. N. (2018). BHEEM: A blockchain-based framework for securing electronic
    health records. *IEEE Globecom Workshops*.
Jianqing Fan, F. H. (2014). Challenges of big data analysis. *National Science Review*, 293–314.
Jingwei Liu, X. L. (2018). BPDS: A blockchain based privacy-preserving data sharing for
    electronic medical records. IEEE.
Jones, D. L., Wagstaff, K., Thompson, D.R., D'Addario, L., Navarro, R., Mattmann, C., Majid,
    W. et al. (2012). Big data challenges for large radio arrays. *Aerospace Conference*
    (pp. 1–6). IEEE.
Joseph Bonneau, A. N. (2014). Mixcoin: Anonymity for bitcoin with accountable mixes.
    *Proceedings of International Conference on Financial Cryptography and Data
    Security* (pp. 486–504). Springer, Berlin, Germany.
Liu, S. (2019). www.statista.com/statistics/647523/worldwide-bitcoin-blockchain-size/.
    Retrieved from www.statista.com.
Madeira, A. (2016). www.cryptocompare.com/coins/guides/what-is-the-block-size-limit/.
    Retrieved from www.cryptocompare.com.
Mardis, E. R. (2016). The challenges of big data. *Disease Models & Mechanisms*, 9, 483–485.
Marín-Marín, J. A., López-Belmonte, J.., Fernández-Campoy, J. M., and Romero-Rodríguez,
    J. M. (2019). Big data in education. A bibliometric review. *Social Sciences—Open
    Access Journal*, 8(8), 1–13.
Marquez, G. (2017). The history of electronic health records (EHRs). Retrieved from www.
    elationhealth.com;www.elationhealth.com/clinical-ehr-blog/history-ehrs/
Maxewell, G. (2013). https://bitcointalk.org/index.php?topic=279249.msg2983902#msg2983902.
    Retrieved from https://bitcointalk.org.
Merisalo, L. (2012). *Protecting Patient Privacy*. Healthcare Registration.
Mian Zhang, Y. J. (2018). Blockchain for healthcare records: A data Prospective. *PeerJ
    Preprints*.
Minal Thakkar, D. C. (2006). Risks, barriers, and benefits of EHR systems: A comparative
    study based on size of hospital. *Perspectives in Health Information Management*.
Möser, M. (2013). Anonymity of bitcoin transactions: An analysis of mixing services.
    Proceedings of Münster Bitcoin Conference (pp. 17–18). Münster, Germany.
Qi Xia, E. B. (2017). BBDS: Blockchain-based data sharing for electronic medical records in
    cloud environments.
Ram Milan, K. K. (2018). Security and privacy challenges in big data. *National Conference
    on Data Anlytics, Machine Learning and Security*, Bilaspur, India.
Reisman, M. (2017). EHRs: The challenge of making electronic data usable and interoperable.
    A Peer-Reviewed Journal for Managed Care and Hospital Formulary Management,
    42, 572–575.
Roman, L. (2009). Combined EMR, EHR and PHR Manage Data for better health. Trade
    Publication.

Ruffing, T., Moreno-Sanchez, P., and Kate, A. (2014). Coinshuffle: Practical decentralized coin mixing. *Proceedings of European Symposium on Research in Computer Security* (pp. 345–364).

Rui Guo, H. S. (2018). Secure attribute-based signature scheme with multiple authorities for blockchain in electronic health records systems. *IEEE Access*, 11676–11686.

Siddiqui, I. (2018). Solution to extract data from ethereum. Retrieved from https://medium.com; https://medium.com/coinmonks/solution-to-extract-data-from-ethereum-52d0b8007d1b.

Statista Research Department. (2016). Internet of Things–number of connected devices worldwide 2015–2025. Retrieved from www.statista.com;www.statista.com/statistics/471264/iot-number-of-connected-devices-worldwide/

Tom Seymour, D. F. (2012). Electronic health records (EHR). *American Journal of Health Sciences*, 3, 201–210.

Vedpal, A., Jain, V., and Bhatnagar, N. (2012). Four C's in supply chain management: Research issues and challenges. *Third International Conference on Emerging Applications of Information Technology (EAIT)* (pp. 264–267). IEEE.

Wolfert, S., Ge, L., Verdouw, C., and Bogaardt, M. J. (2017). Big data in smart farming–A review. *Agricultural Systems*, 69–80.

Zheng, Z., Xie, S., Dai, H. N., Chen, X., and Wang, H. (2018). Blockchain challenges and opportunities: A survey. *International Journal of Web and Grid Services*, 352–375.

10 Test-Driven Development Based on Your Own Blockchain Creation Using Javascript

P. Balakrishnan and L. Ramanathan

CONTENTS

10.1 INTRODUCTION

The monetary transactions among the different entities are often validated and approved by a centralized third-party institution. For instance, the payment of bills or digital transfer of currency definitely demands either a bank or a payment gateway as a third party to accomplish the transaction (Nagasubramanian, 2020) Also, the banks or payment gateways charge transaction fees to the customer to complete the transaction. A similar scenario is extended to different online services, like online purchasing, ticket booking, online gaming, and music services.

In concise, the traditional transaction-based systems voluntarily involve a centralized third party to validate and approve the transactions besides two typical entities involved in the transaction. Contrastingly, blockchain is a technology that uses a decentralized peer-to-peer network to accomplish the transactions between participants without any centralized entity. It is continuously gaining momentum from both the academic communities as well as corporate sectors owing to its prominent features, which include auditability, immutability, decentralization and anonymity. It is a distributed ledger shared among all the nodes in the P2P network (Kumar and Gayathri 2016). Before adding any transactions into that shared ledger, each transaction is verified and validated by the consensus algorithm implemented in the nodes of peer-to-peer network. Upon successful verification, these validated transactions are bundled together and stored into the ledger as blocks, which are later chained together using the hashes of each block. So the blockchain is nothing but chaining of blocks where blocks are used to store the bundle of records/transactions. Each block contains (refer to Figure 10.1) hash code of that block, hash code of previous block, original data together with timestamp. Since each block is linked with each other by storing the hash value of its previous block, any attempt to modify a block leads to change in the hash value of that block which demands the update of hash value in the successive blocks till the end of the blockchain otherwise make that blockchain unusable which makes them more secure. Always, the head of the blockchain is named as genesis block. Genesis block is a dummy block, which contains random values to which the first transaction block points to and takes the hash of genesis block as last-hash in order to create a chain. Miners are specific group of nodes in the P2P network who do the job of validating and verifying the transactions using a unique consensus algorithm. Proof of work (PoW) is a kind of consensus algorithm by which the member nodes construct a trustworthy network atop of untrusted members. The newly generated transaction requests are broadcasted into this P2P network which is subsequently validated by miners thereby eliminating the presence of centralized control. Every block in the blockchain refers the preceding block's hash value thereby ensuring the immutability of the blockchain. For instance, any adversary is trying to tamper the content inside any block leads to hash mismatches in the subsequent block since there is a mismatch between previous hash value field and hash of the modified block. Besides, the transactions are cryptographically sealed using public/private keys. This improvises the anonymity of the blockchain users since the original identity of its users still not traced by the miners. However, the blockchain technology poses the following technical challenges and limitations:

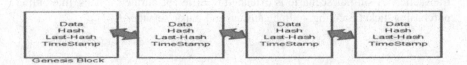

FIGURE 10.1   Representation of blockchain.

Latency: The blockchain technology takes approximately ten minutes to complete one dealing. This is due to the fact that it tries to improvise the security, thereby avoiding the double-spending attacks. In double-spending, a dishonest member is trying to successfully spend the same coin in two different transactions. The double-spending in Bitcoin can easily be identified by verifying and validating the input or consumed coins in a transaction is not spent in any of its previous transactions. Since the above-mentioned procedure is cumbersome and incurs lot of latencies in blockchain network, the modern day blockchain architecture expected to accomplish the new block creation and its verification in seconds without compromising the blockchain security.

Usability: Due to the practical difficulties in the present-day blockchain APIs while developing services, there is a need to propose a more sophisticated developer-friendly blockchain API.

10.1.1 SETTING UP THE ENVIRONMENT

The first step in the development of developer-friendly blockchain API is to prepare the environment. To develop a blockchain project, open the command prompt in Windows 10 and create a directory named **blockchain**. Then create a directory named **dev** inside it, and create two files named **blockchain.js** and **test.js** as shown in Figure 10.2.

After that, execute the command **npminit** from blockchain folder to create a project. This operation demands several inputs from the user, just press the Enter key thereby accepting the default options as shown in Figure 10.3.

At the end of the npm command execution, a json file named package.json is created which is going to maintain out project and its dependencies. The knowledge about the javascript **constructorfunction** and **prototype object** is an elementary step to understand the structure of the blockchain. The constructor function is used

```
E:\blockchain\dev>dir
 Volume in drive E is New Volume
 Volume Serial Number is DCC9-3D25

 Directory of E:\blockchain\dev

27-08-2019  23:53    <DIR>          .
27-08-2019  23:53    <DIR>          ..
27-08-2019  23:41                 2 blockchain.js
27-08-2019  23:38                 0 test.js
               2 File(s)              2 bytes
               2 Dir(s)  271,722,860,544 bytes free
```

FIGURE 10.2 Blockchain project creation.

```
E:\blockchain>npm init
This utility will walk you through creating a package.json file.
It only covers the most common items, and tries to guess sensible defaults.

See 'npm help json' for definitive documentation on these fields
and exactly what they do.

Use 'npm install <pkg>' afterwards to install a package and
save it as a dependency in the package.json file.

Press ^C at any time to quit.
package name: (blockchain)
version: (1.0.0)
description:
entry point: (index.js)
test command:
git repository:
keywords:
author:
license: (ISC)
About to write to E:\blockchain\package.json:

{
  "name": "blockchain",
  "version": "1.0.0",
  "description": "",
  "main": "index.js",
  "scripts": {
    "test": "echo \"Error: no test specified\" && exit 1"
  },
  "author": "",
  "license": "ISC"
}

Is this OK? (yes)
```

FIGURE 10.3   npm command execution.

to create objects, and it can be either a parameterized or non-parameterized one. The following code snippet is used to create a constructor for an Employee class thereby creating multiple Employee objects.

From Figure 10.4, it is evident that employee1 is an object of Employee class, which is having an ID of ID45, firstName of Smith, lastName of Jr and designation of Manager. Besides, the familiarity with Prototype object is also essential to understand the blockchain object. The Prototype object is a reference object that will be referred by several other objects to retrieve any information or functionality. Also, it is possible to extend the properties of the object using Prototype object. In Figure 10.5, it first prints out the default members of the Employee class; then the Prototype object is included from the Employee class. Subsequently, the Employee class object is extended with **location** attribute whose value is set as **India**. However, the extended feature is not reflected in the Employee class object. Then a getter method is designed to get back the values stored in the location attribute.

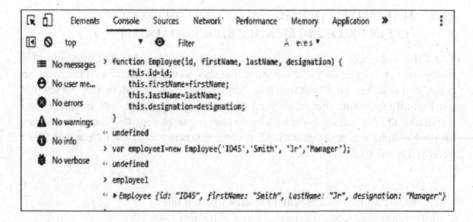

FIGURE 10.4 Employee class creation using employee constructor.

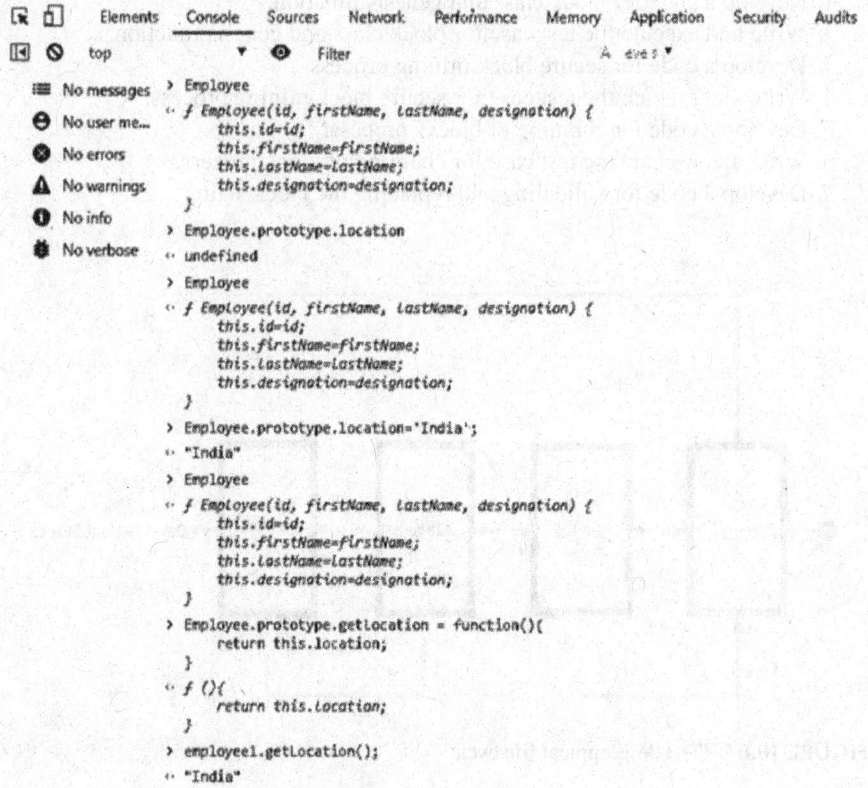

FIGURE 10.5 Prototype object implementation on employee class object.

10.2   TEST-DRIVEN DEVELOPMENT (TDD) APPROACH FOR DEVELOPER-FRIENDLY BLOCKCHAIN API

The TDD is an approach to develop software that implements the repetition of a really short development cycle meaning that the developer needs to implement an automated test case for a new function then develops a new code which passes this test partially following that refactoring of the newly developed code to comply the standards. This test-develop-refactor cycle is going to be repeated for several times till the blockchain is developed. Generally, the following steps are followed in TDD (refer to Figure 10.6):

1. Develop a test case.
2. Execute all the test cases and identify where it fails.
3. Develop some code that passes some of the test cases.
4. Refactor the code to comply with the chosen standards.
5. Repeat the above steps.

The application of TDD for the development of blockchain API are as follows:

1. Develop a code for block class and genesis function.
2. Write and execute the test case for block class and genesis function.
3. Develop a code for secure block mining process.
4. Write and execute the test case for secure block mining process.
5. Develop a code for chaining of blocks process.
6. Write and execute the test case for chaining of block process.
7. Develop a code for validating and replacing the blockchain.

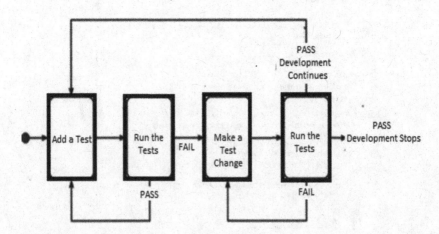

FIGURE 10.6   TDD development life cycle.

8. Write and execute the test case for validating and replacing the blockchain.
9. Develop a code for dynamically adjusting the difficulty level in proof of work (PoW).
10. Write and execute the test case for adjusting the difficulty level in PoW.

10.2.1 Designing of Block Class

The blocks are the most fundamental unit of the blockchain, which are needed to create a chain. At the cores, a block consists of six pieces of information such as timestamp, hash of the block, hash of the previous block, data, nonce, and difficulty.

Timestamp: A timestamp is nothing but the time at which the block is created. It is created as javaScript date object, which will help to record the time down to millisecond.

Hash: Hash is defined as the string of characters that is created on the basis on the unique data and the combination of other pieces of data. Even the slightest change in the block will lead to the change of the hash value. Therefore, it helps to create the obstacle for the attack by any malicious users.

Last-Hash: Last-Hash is the hash value of the previous block. The blockchain can be defined as the list on linked block, with the help of Last-Hash value of the link is created among the blocks. In short, if the hash value of previous block and the last-hash value of current block do not match, then that link will be deleted.

Data: Data column is the information of that needs to be stored in the block. It is composed of any data-type. In the current product, the data is the combination of data-type with valid information about the sender, receiver and the amount.

Nonce (Number only used once): It is a 4-byte field that is modified by the miners of the network to ensure the hash of the block below the target which is a 256 bit number of the network.

Difficulty: It represents the degree of challenge incurred while finding the hash value below a given target. For instance, the lower the target, the higher the difficulty and vice versa.

The origin of the blockchain is known as genesis block. It is hard coded with the dummy values of timestamp, hash, last-hash and the data, in order to start the blockchain. This way the first block of the blockchain will take the last-hash value of genesis block.

The following piece of code represents the creation of block class contains the genesis block function with hard coded global values:

LISTING 10.1   CODE FOR BLOCK CLASS
WITH GENESIS BLOCK

```
class Block {
constructor({ timestamp, lastHash, hash, data, nonce,
difficulty }) {
this.timestamp = timestamp;
this.lastHash = lastHash;
this.hash = hash;
this.data = data;
this.nonce = nonce;
this.difficulty = difficulty;
  }
static genesis() {
returnnew this(GENESIS_DATA);
  }
const GENESIS_DATA = {
timestamp: 1,
lastHash: '-----',
hash: 'hash-one',
difficulty: INITIAL_DIFFICULTY,
nonce: 0,
data: []
};
```

As in influenced by the test-driven development approach, the test case for the block class is designed to check whether the core four pieces of the block has the valid inputs:

LISTING 10.2   TEST CASES FOR BLOCK
CLASS AND THE GENESIS FUNCTION

```
it('has a timestamp, lastHash, hash, and data property',
  () => {
expect(block.timestamp).toEqual(timestamp);
expect(block.lastHash).toEqual(lastHash);
expect(block.hash).toEqual(hash);
expect(block.data).toEqual(data);
expect(block.nonce).toEqual(nonce);
expect(block.difficulty).toEqual(difficulty);
  });
describe('genesis()', () => {
```

(Continued)

LISTING 10.2 (Continued)   TEST CASES FOR BLOCK
CLASS AND THE GENESIS FUNCTION

```
constgenesisBlock = Block.genesis();
it('returns a Block instance', () => {
expect(genesisBlockinstanceof Block).toBe(true);
  });
it('returns the genesis data', () => {
expect(genesisBlock).toEqual(GENESIS_DATA);
  });
  });
```

10.2.2 Designing of Block Mining Process

Mining is a process of bundling the verified transactions as a block and adding it to the previous block in the shared ledger by chaining them using their hash values. However, the mining process can be applied over the blocks that are satisfying the following conditions:

1. The block can be added with other blocks in the shared ledger if and only if it is an instance of block class.
2. The last-hash of the present block should be same as the hash value of the preceding block.
3. The block should contain some data attributes that are going to be shared.
4. The block should contain a timestamp.

The following code represents the mining function named mineBlock with lastBlock and data as the arguments.

LISTING 10.3   CODE FOR BLOCK MINING PROCESS

```
staticmineBlock({ lastBlock, data }) {
constlastHash = lastBlock.hash;
let hash, timestamp;
let { difficulty } = lastBlock;
let nonce = 0;
do {
nonce++;
timestamp = Date.now();
difficulty = Block.adjustDifficulty({ originalBlock:
    lastBlock, timestamp });
```

(Continued)

LISTING 10.3 (Continued)   CODE FOR BLOCK MINING PROCESS

```
hash = cryptoHash(timestamp, lastHash, data, nonce,
difficulty);
    } while (hexToBinary(hash).substring(0, difficulty)
    !== '0'.repeat(difficulty));
returnnew this({ timestamp, lastHash, data, difficulty,
    nonce, hash });
  }
```

Subsequently, the test-case for the mineBlockfunction is developed and given in Listing 10.4.

LISTING 10.4   TEST-CASE FOR MINEBLOCK FUNCTION

```
describe('mineBlock()', () => {
constlastBlock = Block.genesis();
const data = 'mined data';
constminedBlock = Block.mineBlock({ lastBlock, data });
it('returns a Block instance', () => {
expect(minedBlockinstanceof Block).toBe(true);
  });
it('sets the `lastHash` to be the `hash` of the
    lastBlock', () => {
expect(minedBlock.lastHash).toEqual(lastBlock.hash);
  });
it('sets the `data`', () => {
expect(minedBlock.data).toEqual(data);
  });
it('sets a `timestamp`', () => {
expect(minedBlock.timestamp).not.toEqual(undefined);
  });
  });
```

Apart from that, the blockchain becomes immutable meaning that its contents cannot be modified because of the hashing algorithms such as, SHA-256. The crypto is a native module of javaScript which is used in encryption. The cryptoHash constant is combining all the values in the block, create the hash value with 256 bits using SHA-256 algorithm, convert the hash value into a Hex format with 64 characters and returns it. The code for the hashing of block contents is given in Listing 10.5.

**LISTING 10.5   CODE FOR HASHING THE
CONTENTS OF THE BLOCK**

```
describe('mineBlock()', () => {
it('creates a SHA-256 `hash` based on the proper inputs',
    () => {
expect(minedBlock.hash) .toEqual(cryptoHash(
minedBlock.timestamp, minedBlock.nonce,  minedBlock.
    difficulty, lastBlock.hash, data ) ); });
```

The test-case for the cryptoHash function with SHA-256 hashing algorithm is given in Listing 10.6.

**LISTING 10.6   TEST-CASE FOR CRYPTOHASH
FUNCTION WITH SHA-256 HASHING ALGORITHM**

```
The test for the crypto-Hash module:
constcryptoHash = require('./crypto-hash');
describe('cryptoHash()', () => {
it('generates a SHA-256 hashed output', () => {
expect(cryptoHash('foo'))
    .toEqual('b2213295d564916f89a6a42455567c87c3f480fcd7a1
        c15e220f17d7169a790b');
 });
```

10.2.3   CHAINING OF BLOCKS

After that, the secured blocks that are created from the previous section need to be chained together to formulate the blockchain. Naturally, the first block of the chain is always a genesis block, which should have the capability to add other blocks using addBlock() function. The code for blockchain is given in index.js which is given in Listing 10.7.

LISTING 10.7   CODE FOR BLOCKCHAIN

```
const Block = require('./block');
classBlockchain {
constructor() {
this.chain = [Block.genesis()];
 }

addBlock({ data }) {
constnewBlock = Block.mineBlock({
lastBlock: this.chain[this.chain.length-1],
data
 });
this.chain.push(newBlock);
 }
```

Following that, the test case for blockchain is given in index.test.js file. Firstly, the blockchain object is checked whether it is an instance of array. Later, the first block of blockchain is extracted and tested whether it is a genesis block.

LISTING 10.8   TEST-CASE FOR BLOCKCHAIN

```
constBlockchain = require('./index');
const Block = require('./block');
describe('Blockchain', () => {
it('contains a `chain` Array instance', () => {
        expect(blockchain.chaininstanceof Array).toBe(true);
        });
it('starts with the genesis block', () => {
expect(blockchain.chain[0]).toEqual(Block.genesis());
 });
it('adds a new block to the chain', () => {
        constnewData = 'foo bar';
        blockchain.addBlock({ data: newData });
expect(blockchain.chain[blockchain.chain.length-1].data).
    toEqual(newData);
});
```

10.2.4   BLOCKCHAIN VALIDATION AND REPLACEMENT

The blockchain validation is process of verifying whether the blocks in the blockchain are constructed properly by inspecting their block fields, hash and last-hash references, which are implemented in isValidChain() function, which is outlined in Listing 10.9. The test-case function for this blockchain validation should return false if any of the above-mentioned inspections fail, which is given in Listing 10.10.

**LISTING 10.9   CODE FOR VALIDATING
THE BLOCKS IN A BLOCKCHAIN**

```
staticisValidChain(chain){
if (JSON.stringify(chain[0]) !== JSON.stringify(Block.
   genesis())) {
returnfalse
   };

for (let i=1; i<chain.length; i++) {
const { timestamp, lastHash, hash, nonce, difficulty,
   data } = chain[i];
constactualLastHash = chain[i-1].hash;
constlastDifficulty = chain[i-1].difficulty;

if (lastHash !== actualLastHash) returnfalse;

constvalidatedHash = cryptoHash(timestamp, lastHash,
   data, nonce, difficulty);
if (hash !== validatedHash) returnfalse;
if (Math.abs(lastDifficulty - difficulty) >1) returnfalse;
   }
returntrue;
   }
```

**LISTING 10.10   TEST-CASE FOR VALIDATING
THE BLOCKS IN THE BLOCKCHAIN**

```
describe('isValidChain()', () => {
describe('when the chain does not start with the genesis
block', () => {
it('returns false', () => {
blockchain.chain[0] = { data: 'fake-genesis' };

expect(Blockchain.isValidChain(blockchain.chain)).toBe(false);
   });
   });
describe('when the chain starts with the genesis block
   and has multiple blocks', () => {
beforeEach(() => {
blockchain.addBlock({ data: 'Bears' });
blockchain.addBlock({ data: 'Beets' });
blockchain.addBlock({ data: 'BattlestarGalactica' });
   });
describe('and a lastHash reference has changed', () => {
it('returns false', () => {
blockchain.chain[2].
lastHash = 'broken-lastHash';
```

(Continued)

**LISTING 10.10 (Continued)   TEST-CASE FOR
VALIDATING THE BLOCKS IN THE BLOCKCHAIN**

```
expect(Blockchain.isValidChain(blockchain.chain)).
   toBe(false);
   });
   });
describe('and the chain contains a block with an invalid
   field', () => {
it('returns false', () => {
blockchain.chain[2].data = 'some-bad-and-evil-data';

expect(Blockchain.isValidChain(blockchain.chain)).
   toBe(false);
   });
   });
```

Subsequently, the blockchain replacement process, which replaces an array of blocks in a blockchain with a new blockchain as long as the new chain is longer and the incoming chain of blocks are valid. The code for blockchain replacement and the corresponding test-cases for blockchain replacement are given in Listing 10.11 and Listing 10.12, respectively.

LISTING 10.11   CODE FOR BLOCKCHAIN REPLACEMENT

```
replaceChain(chain, validateTransactions, onSuccess) {
if (chain.length<= this.chain.length) {
console.error('The incoming chain must be longer');
return;
   }

if (!Blockchain.isValidChain(chain)) {
console.error('The incoming chain must be valid');
return;
   }

if (validateTransactions&& !this.
   validTransactionData({ chain })) {
console.error('The incoming chain has invalid data');
return;
   }

if (onSuccess) onSuccess();
console.log('replacing chain with', chain);
this.chain = chain;
   }
```

LISTING 10.12   TEST-CASE FOR REPLACEMENT OF BLOCKCHAIN

```
describe('replaceChain()', () => {
letlogMock;

beforeEach(() => {
logMock = jest.fn();

  global.console.log = logMock;
 });

describe('when the new chain is not longer', () => {
beforeEach(() => {
newChain.chain[0] = { new: 'chain' };

blockchain.replaceChain(newChain.chain);
  });

it('does not replace the chain', () => {
expect(blockchain.chain).toEqual(originalChain);
  });

it('logs an error', () => {
expect(errorMock).toHaveBeenCalled();
  });
 });

describe('when the new chain is longer', () => {
beforeEach(() => {
newChain.addBlock({ data: 'Bears' });
newChain.addBlock({ data: 'Beets' });
newChain.addBlock({ data: 'BattlestarGalactica' });
  });

describe('and the chain is invalid', () => {
beforeEach(() => {
newChain.chain[2].hash = 'some-fake-hash';

blockchain.replaceChain(newChain.chain);
    });

it('does not replace the chain', () => {
expect(blockchain.chain).toEqual(originalChain);
  });

it('logs an error', () => {
expect(errorMock).toHaveBeenCalled();
  });
 });
```

(Continued)

**LISTING 10.12 (Continued)   TEST-CASE FOR
REPLACEMENT OF BLOCKCHAIN**

```
describe('and the chain is valid', () => {
beforeEach(() => {
blockchain.replaceChain(newChain.chain);
  });

it('replaces the chain', () => {
expect(blockchain.chain).toEqual(newChain.chain);
  });

it('logs about the chain replacement', () => {
expect(logMock).toHaveBeenCalled();
  });
  });
  });
```

10.2.5 Proof of Work (PoW) Process

The primary objective of the PoW is to establish a trust on the participating nodes, which are in an untrusted environment. The participating nodes are requested to devote a certain amount of time in carrying out compute-intensive work in order to append new blocks into the blockchain rather than allowing anyone to freely add a block to data. The immediate benefit is that this deters attackers from trying to rewrite the entire block in history with corrupt and invalid data. It is expensive for the attackers to spend the resources in order to corrupt the data but still reasonable for the honest contributors to spend the time and money to add a block because in decentralized blockchain network any node has a capacity to submit a chain to the network, adding one block is not such an expensive task as long as a block is valid and based on the existing valid chain. If the malicious user pays attention to corrupt the data of one block, the hash value will change and it will subsequently break the chain as the hash value of previous block should be equal to last hash value of the current block. Hence, it will be really expensive for the attacker to change the entire chain. For that, a hashcash-based PoW is implemented in this API wherein the difficulty will be adjusted dynamically to modify the complexity of mining process. Hence, the code and the test case for the difficulty adjustment are given in Listing 10.13 and Listing 10.14, respectively.

LISTING 10.13   CODE FOR DYNAMIC DIFFICULTY ADJUSTMENT FOR PoW

```
do {
nonce++;
timestamp = Date.now();
difficulty = Block.adjustDifficulty({ originalBlock:
   lastBlock, timestamp });
hash = cryptoHash(timestamp, lastHash, data, nonce,
   difficulty);
 } while (hexToBinary(hash).substring(0, difficulty)
   !== '0'.repeat(difficulty));

returnnew this({ timestamp, lastHash, data, difficulty,·
   nonce, hash });
 }

staticadjustDifficulty({ originalBlock, timestamp }) {
const { difficulty } = originalBlock;

if (difficulty <1)
        return1;

if ((timestamp - originalBlock.timestamp) > MINE_RATE )
        return difficulty - 1;

return difficulty + 1;
 }
```

LISTING 10.14   TEST-CASE FOR CHECKING THE DIFFICULTY LEVEL IN PoW

```
describe('adjustDifficulty()', () => {
it('raises the difficulty for a quickly mined block', ()
   => {
expect(Block.adjustDifficulty({
originalBlock: block, timestamp: block.
   timestamp + MINE_RATE - 100
   })).toEqual(block.difficulty+1);
 });

it('lowers the difficulty for a slowly mined block', ()
   => {
expect(Block.adjustDifficulty({
```

(Continued)

```
originalBlock: block, timestamp: block.
   timestamp + MINE_RATE + 100
  })).toEqual(block.difficulty-1);
 });

it('has a lower limit of 1', () => {
block.difficulty = -1;

expect(Block.adjustDifficulty({ originalBlock: block })).
   toEqual(1);
  });
 });
```

10.3   DEVELOPMENT OF REST-BASED APIs FOR BLOCKCHAIN APPLICATIONS

As the core logics for implementing a blockchain is completed, it is essential to incorporate the additional capabilities on top of this core logic. Alternatively, the primary objective is to formulate a network of nodes that are running blockchain applications, which can be interacted with each other through a common API. Hence, each node in the network is going to run the same API in order to expose common methods that can be used to read each other's information. The Express server is used to implement various HTTP post and get requests. The core concept of API is to support an overall blockchain network. The HTTP GET and POST requests are given in Listing 10.15 and Listing 10.16, repectively.

LISTING 10.15   HTTP GET REQUESTS FOR BLOCKCHAIN API

```
app.get('/api/blocks', (req, res) => {
res.json(blockchain.chain);
});
app.get('/api/blocks/length', (req, res) => {
res.json(blockchain.chain.length);
});
app.get('/api/blocks/:id', (req, res) => {
const { id } = req.params;
```

(Continued)

**LISTING 10.15 (Continued)   HTTP GET
REQUESTS FOR BLOCKCHAIN API**

```
const { length } = blockchain.chain;
constblocksReversed = blockchain.chain.slice().reverse();
letstartIndex = (id-1) * 5;
letendIndex = id * 5;
startIndex = startIndex< length ? startIndex : length;
endIndex = endIndex< length ? endIndex : length;
res.json(blocksReversed.slice(startIndex, endIndex));
});
app.get('/api/transaction-pool-map', (req, res) => {
res.json(transactionPool.transactionMap);
});
app.get('/api/mine-transactions', (req, res) => {
transactionMiner.mineTransactions();
res.redirect('/api/blocks');
});
app.get('/api/wallet-info', (req, res) => {
const address = wallet.publicKey;
res.json({
address,
balance: Wallet.calculateBalance({ chain: blockchain.
   chain, address })
 });
});
app.get('/api/known-addresses', (req, res) => {
constaddressMap = {};
for (let block ofblockchain.chain) {
for (let transaction ofblock.data) {
const recipient = Object.keys(transaction.outputMap);
recipient.forEach(recipient =>addressMap[recipient]
   = recipient);
   }
 }
res.json(Object.keys(addressMap));
});
app.get('*', (req, res) => {
res.sendFile(path.join(__dirname, 'client/dist/index.
   html'));
});
```

LISTING 10.16   HTTP POST REQUEST FOR BLOCKCHAIN API

```
app.post('/api/mine', (req, res) => {
const { data } = req.body;
blockchain.addBlock({ data });
pubsub.broadcastChain();
res.redirect('/api/blocks');
});
app.post('/api/transact', (req, res) => {
const { amount, recipient } = req.body;
let transaction = transactionPool
    .existingTransaction({ inputAddress: wallet.
    publicKey });
try {
if (transaction) {
transaction.update({ senderWallet: wallet, recipient,
    amount });
    } else {
transaction = wallet.createTransaction({
recipient,
amount,
chain: blockchain.chain
    });
    }
  } catch(error) {
returnres.status(400).json({ type: 'error', message:
    error.message });
  }
transactionPool.setTransaction(transaction);
pubsub.broadcastTransaction(transaction);
res.json({ type: 'success', transaction });
});
constsyncWithRootState = () => {
request({ url: `${ROOT_NODE_ADDRESS}/api/blocks` },
    (error, response, body) => {
if (!error &&response.statusCode === 200) {
constrootChain = JSON.parse(body);
console.log('replace chain on a sync with', rootChain);
blockchain.replaceChain(rootChain);
  }
 });
request({ url: `${ROOT_NODE_ADDRESS}/api/transaction-
    pool-map` }, (error, response, body) => {
if (!error &&response.statusCode === 200) {
constrootTransactionPoolMap = JSON.parse(body);
console.log('replace transaction pool map on a sync
    with', rootTransactionPoolMap);
```

(Continued)

**LISTING 10.16 (Continued)   HTTP
POST REQUEST FOR BLOCKCHAIN API**

```
transactionPool.setMap(rootTransactionPoolMap);
  }
});
};
if (isDevelopment) {
constwalletFoo = new Wallet();
constwalletBar = new Wallet();
constgenerateWalletTransaction = ({ wallet, recipient,
   amount }) => {
const transaction = wallet.createTransaction({
recipient, amount, chain: blockchain.chain
  });
transactionPool.setTransaction(transaction);
};
constwalletAction = () =>generateWalletTransaction({
wallet, recipient: walletFoo.publicKey, amount: 5
  });
constwalletFooAction = () =>generateWalletTransaction({
wallet: walletFoo, recipient: walletBar.publicKey,
   amount: 10
  });
constwalletBarAction = () =>generateWalletTransaction({
wallet: walletBar, recipient: wallet.publicKey,
   amount: 15
  });
for (let i=0; i<20; i++) {
if (i%3 === 0) {
walletAction();
walletFooAction();
  } elseif (i%3 === 1) {
walletAction();
walletBarAction();
  } else {
walletFooAction();
walletBarAction();
  }
transactionMiner.mineTransactions();
  }
}
let PEER_PORT;
if (process.env.GENERATE_PEER_PORT === 'true') {
 PEER_PORT = DEFAULT_PORT + Math.ceil(Math.
   random() * 1000);
}
```

(Continued)

**LISTING 10.16 (Continued)   HTTP
POST REQUEST FOR BLOCKCHAIN API**

```
const PORT = process.env.PORT || PEER_PORT || DEFAULT_PORT;
app.listen(PORT, () => {
console.log(`listening at localhost:${PORT}`);
if (PORT !== DEFAULT_PORT) {
syncWithRootState();
}
```

10.4 WALLETS IN CRYPTOCURRENCY

A wallet in cryptocurrency may be a hardware device, medium, software or a service that stocks the key pairs that will be used to verify the rights, credit or debit the cryptocurrencies. Here, the public key of a wallet may be used to credit the cryptocurrencies from any other wallet whereas the private key is used to debit cryptocurrency from this wallet to any other wallet. The code and test-case for transactions using a wallet is given in Listing 10.17 and Listing 10.18, respectively.

LISTING 10.17   CODE FOR TRANSACTIONS USING A WALLET

```
class Wallet {
constructor() {
this.balance = STARTING_BALANCE;
this.keyPair = ec.genKeyPair();
this.publicKey = this.keyPair.getPublic().encode('hex');
}

sign(data) {
returnthis.keyPair.sign(cryptoHash(data))
}

createTransaction({ recipient, amount, chain }) {
if (chain) {
```

(Continued)

LISTING 10.17 (Continued)   CODE FOR TRANSACTIONS
USING A WALLET

```
this.balance = Wallet.calculateBalance({
chain,
address: this.publicKey
    });
    }


if (amount >this.balance) {
thrownew Error('Amount exceeds balance');
    }
returnnew Transaction({ senderWallet: this, recipient,
    amount });
    }


staticcalculateBalance({ chain, address }) {
lethasConductedTransaction = false;
letoutputsTotal = 0;


for (let i=chain.length-1; i>0; i--) {
const block = chain[i];


for (let transaction ofblock.data) {
if (transaction.input.address === address) {
hasConductedTransaction = true;
    }


constaddressOutput = transaction.outputMap[address];


if (addressOutput) {
outputsTotal = outputsTotal + addressOutput;
    }
    }


if (hasConductedTransaction) {
break;
    }
    }


returnhasConductedTransaction ? outputsTotal : STARTING_
    BALANCE + outputsTotal;
    }
};
```

LISTING 10.18 TEST-CASE FOR WALLET

```
describe('Wallet', () => {
let wallet;

beforeEach(() => {
wallet = new Wallet();
 });

it('has a `balance`', () => {
expect(wallet).toHaveProperty('balance');
 });

it('has a `publicKey`', () => {
expect(wallet).toHaveProperty('publicKey');
 });

describe('signing data', () => {
const data = 'foobar';

it('verifies a signature', () => {
expect(
verifySignature({
publicKey: wallet.publicKey,
data,
signature: wallet.sign(data)
    })
   ).toBe(true);
  });

it('does not verify an invalid signature', () => {
expect(
verifySignature({
publicKey: wallet.publicKey,
data,
signature: new Wallet().sign(data)
    })
   ).toBe(false);
  });
 });

describe('createTransaction()', () => {
describe('and the amount exceeds the balance', () => {
it('throws an error', () => {
expect(() =>wallet.createTransaction({ amount: 999999,
   recipient: 'foo-recipient' }))
```

(Continued)

LISTING 10.18 (Continued)   TEST-CASE FOR WALLET

```
    .toThrow('Amount exceeds balance');
  });
  });

describe('and the amount is valid', () => {
let transaction, amount, recipient;

beforeEach(() => {
amount = 50;
recipient = 'foo-recipient';
transaction = wallet.
createTransaction({ amount, recipient });
    });

it('creates an instance of `Transaction`', () => {
expect(transaction instanceof Transaction).toBe(true);
    });

it('matches the transaction input with the wallet', () => {
expect(transaction.input.address).toEqual(wallet.
   publicKey);
      });

it('outputs the amount the recipient', () => {
expect(transaction.outputMap[recipient]).toEqual(amount);
   });
  });

describe('and a chain is passed', () => {
it('calls `Wallet.calculateBalance`', () => {
constcalculateBalanceMock = jest.fn();

constoriginalCalculateBalance = Wallet.calculateBalance;

Wallet.calculateBalance = calculateBalanceMock;

wallet.createTransaction({
recipient: 'foo',
amount: 10,
chain: newBlockchain().chain
   });

expect(calculateBalanceMock).toHaveBeenCalled();
```

(Continued)

LISTING 10.18 (Continued)   TEST-CASE FOR WALLET

```
Wallet.calculateBalance = originalCalculateBalance;
    });
   });
  });

describe('calculateBalance()', () => {
letblockchain;

beforeEach(() => {
blockchain = newBlockchain();
  });

describe('and there are no outputs for the wallet', () => {
it('returns the `STARTING_BALANCE`', () => {
expect(
Wallet.calculateBalance({
chain: blockchain.chain,
address: wallet.publicKey
    })
   ).toEqual(STARTING_BALANCE);
  });
 });

describe('and there are outputs for the wallet', () => {
lettransactionOne, transactionTwo;

beforeEach(() => {
transactionOne = new Wallet().createTransaction({
recipient: wallet.publicKey,
amount: 50
   });

transactionTwo = new Wallet().createTransaction({
recipient: wallet.publicKey,
amount: 60
   });

blockchain.addBlock({ data: [transactionOne,
   transactionTwo] });
  });

it('adds the sum of all outputs to the wallet balance',
() => {
expect(
Wallet.calculateBalance({
```

(Continued)

LISTING 10.18 (Continued)   TEST-CASE FOR WALLET

```
chain: blockchain.chain,
address: wallet.publicKey
    })
  ).toEqual(
     STARTING_BALANCE +
transactionOne.outputMap[wallet.publicKey] +
transactionTwo.outputMap[wallet.publicKey]
    );
    });

describe('and the wallet has made a transaction', () => {
letrecentTransaction;

beforeEach(() => {
recentTransaction = wallet.createTransaction({
recipient: 'foo-address',
amount: 30
    });

blockchain.addBlock({ data: [recentTransaction] });
    });

it('returns the output amount of the recent transaction',
() => {
expect(
Wallet.calculateBalance({
chain: blockchain.chain,
address: wallet.publicKey
    })
    ).toEqual(recentTransaction.outputMap[wallet.
    publicKey]);
    });

describe('and there are outputs next to and after the
recent transaction', () => {
letsameBlockTransaction, nextBlockTransaction;

beforeEach(() => {
recentTransaction = wallet.createTransaction({
recipient: 'later-foo-address',
amount: 60
    });

sameBlockTransaction = Transaction.rewardTransaction({
minerWallet: wallet });
```

(Continued)

```
             LISTING 10.18 (Continued)   TEST-CASE FOR WALLET
blockchain.addBlock({ data: [recentTransaction,
sameBlockTransaction] });

nextBlockTransaction = new Wallet().createTransaction({
recipient: wallet.publicKey, amount: 75
        });

blockchain.addBlock({ data: [nextBlockTransaction] });
      });

it('includes the output amounts in the returned balance',
() => {
expect(
Wallet.calculateBalance({
chain: blockchain.chain,
address: wallet.publicKey
        })
        ).toEqual(
recentTransaction.outputMap[wallet.publicKey] +
sameBlockTransaction.outputMap[wallet.publicKey] +
nextBlockTransaction.outputMap[wallet.publicKey]
        );
      });
    });
   });
  });
 });
});
```

10.4.1 Transaction Pool

The transaction pool is a data structure that is used to collect the transactions that can be implemented as an array or a map with a key value structure. Apart from this, it has to support the following behaviors:

- Collects a unique set of transaction objects
- Supports updating of transactions whenever change has been submitted by a wallet
- Rewrites multiple transactions in the inner collection

This is shown in Listing 10.19.

LISTING 10.19   CODE FOR IMPLEMENTING
TRANSACTION POOL

```
const Transaction = require('./transaction');
classTransactionPool {
constructor() {
this.transactionMap = {};
  }
clear() {
this.transactionMap = {};
  }
setTransaction(transaction) {
this.transactionMap[transaction.id] = transaction;
  }
setMap(transactionMap) {
this.transactionMap = transactionMap;
  }
existingTransaction({ inputAddress }) {
const transactions = Object.values(this.transactionMap);
returntransactions.find(transaction =>transaction.input.
address === inputAddress);
  }
validTransactions() {
returnObject.values(this.transactionMap).filter(
transaction =>Transaction.validTransaction(transaction)
    );
  }
clearBlockchainTransactions({ chain }) {
for (let i=1; i<chain.length; i++) {
const block = chain[i];
for (let transaction of block.data) {
if (this.transactionMap[transaction.id]) {
deletethis.transactionMap[transaction.id];
    } }
   }
  }
}
module.exports = TransactionPool;
```

The tests to validate the code:

LISTING 10.20   TEST-CASE FOR TRANSACTION POOL

```
constTransactionPool = require('./transaction-pool');
const Transaction = require('./transaction');
const Wallet = require('./index');
constBlockchain = require('../blockchain');
describe('TransactionPool', () => {
lettransactionPool, transaction, senderWallet;

beforeEach(() => {
transactionPool = new TransactionPool();
senderWallet = new Wallet();
transaction = new Transaction({
senderWallet,
recipient: 'fake-recipient',
amount: 50
    });
  });
describe('setTransaction()', () => {
it('adds a transaction', () => {
transactionPool.setTransaction(transaction);
expect(transactionPool.transactionMap[transaction.id])
    .toBe(transaction);
    });
  });
describe('existingTransaction()', () => {
it('returns an existing transaction given an input
    address', () => {
transactionPool.setTransaction(transaction);
expect(
transactionPool.existingTransaction({ inputAddress:
    senderWallet.publicKey })
    ).toBe(transaction);
    });
  });
describe('validTransactions()', () => {
letvalidTransactions, errorMock;
beforeEach(() => {
validTransactions = [];
errorMock = jest.fn();
global.console.error = errorMock;
for (let i=0; i<10; i++) {
transaction = new Transaction({
senderWallet,
recipient: 'any-recipient',
amount: 30
    });
```

(Continued)

**LISTING 10.20 (Continued) TEST-CASE FOR
TRANSACTION POOL**

```
if (i%3===0) {
transaction.input.amount = 999999;
    } else if (i%3===1) {
transaction.input.signature = new Wallet().sign('foo');
    } else {
validTransactions.push(transaction);
    }


transactionPool.setTransaction(transaction);
    }
 });
it('returns valid transaction', () => {
expect(transactionPool.validTransactions()).
    toEqual(validTransactions);
    });
it('logs errors for the invalid transactions', () => {
transactionPool.validTransactions();
expect(errorMock).toHaveBeenCalled();
    });
 });
describe('clear()', () => {
it('clears the transactions', () => {
transactionPool.clear();
expect(transactionPool.transactionMap).toEqual({});
    });
 });
describe('clearBlockchainTransactions()', () => {
it('clears the pool of any existing
    blockchain transactions', () => {
constblockchain = new Blockchain();
constexpectedTransactionMap = {};
for (let i=0; i<6; i++) {
const transaction = new Wallet().createTransaction({
recipient: 'foo', amount: 20
    });
transactionPool.setTransaction(transaction);
if (i%2===0) {
blockchain.addBlock({ data: [transaction] })
    } else {
expectedTransactionMap[transaction.id] = transaction;
    }
    }
```

(Continued)

**LISTING 10.20 (Continued)   TEST-CASE FOR
TRANSACTION POOL**

```
transactionPool.clearBlockchainTransactions({ chain:
    blockchain.chain });
expect(transactionPool.transactionMap).toEqual(expectedTr
    ansactionMap);
    });
  });
});
```

10.5   CONCLUSION

Cryptocurrencies are new kind of digital assets that are revolutionizing the manner by which the international markets are associated with each other, thereby ruliong out the obstacles surrounding normative national currencies and exchange rates. Blockchain is a distributed ledger that bundles the transactions as blocks after the successful verification and validation using the consensus algorithm implemented in the nodes of P2P networks, which are later chained together using the hash codes of each block. The cryptocurrencies use the blockchain network to do the transactions in secured way without any centralized authority. This chapter elucidates a TDD-based methodology to create a new blockchain using Javascript. Since the application of blockchain technology is not restricted solely to business, the steps explained can be used in other domains as well.

REFERENCES

Kumar, S. R., & Gayathri, N. (2016). Trust based data transmission mechanism in MANET using sOLSR. In *Annual Convention of the Computer Society of India* (pp. 169–180). Springer, Singapore.
Nagasubramanian, G., Sakthivel, R. K., Patan, R. et al. (2020). Securing e-health records using keyless signature infrastructure blockchain technology in the cloud. *Neural Computing & Applications*, 32, 639–647. https://doi.org/10.1007/s00521-018-3915-1

11 A Study on Privacy-Preserving Models Using Blockchain Technology for IoT

Syed Muzamil Basha, J. Janet, and S. Balakrishnan

CONTENTS

11.1 INTRODUCTION

Transactions cannot be costly and time-consuming. A report from Nielsen estimates that in 2016, the losses topped over $24.71 billion of credit card transactions, and the merchants were still responsible for 28% of those losses, making it a huge and very expensive way of managing their transactions. But based on blockchain technology, these transactions can happen much faster today, and on ramps and off ramps from the Bitcoin or the cryptocurrency to the real world are possible. So for example, you can buy an Amazon gift card or a Starbucks gift card with Bitcoin today. Another area we will see blockchain probably sooner than later is human resources. With blockchain technology, we can track verified credentials, like course completion, awards won, and training certification in an immutable way.

The impact of blockchain technology has on consumers, business models, marketing, and advertising [42]. Now-a-days, everyone can do transaction to every others, and add billions of people to the marketplace, where transactions can be any size. They can be cheap, and the charges imposed on performing such transactions can become zero. And they can be fast, where they are immutable, which means permanent with in-records. It allows new businesses marketing and advertisement opportunities. And it impacts every transaction step, from compliance and supply chain to advertisement, marketing, and sales. Blockchain technology [40,41] is revolutionizing the world economy [45]. Trust is established through peer-to-peer mass collaborations and sophisticated computer code rather than through a powerful central institution, like a bank or a government. Changing transactions and interactions and finance, business, international collaborations, and the government's tool chest are enabled through blockchain technology.

Social media can enable people to manage their identity in the privacy of their data instead of having to rely on centralized entities, such as Google, Facebook, and Twitter when we combine them with blockchain technology. And that will move us from a centralized to decentralized social media, where tracking of customer behavior across websites becomes really hard. The move to blockchain-based social networks that preserve anonymity like Mastodon Systemic, decentralized versions basically of Twitter and Reddit, really change and show what's possible in a world of disintermediation of Twitter and Facebook if they don't disseminate themself first. The data ownership is going to change. Customers' data will belong to organizations locked away in corporate databases today.

Blockchain offers a new opportunity to chain our customers to our brands and marketing. Loyalty programs have been a huge challenge to establish; a lot of people are dissatisfied. But blockchain technology creates the possibility of microtransaction very, very cost-efficient, and therefore, creates new opportunities, collect loyalty points, bringing different companies, different brands together, and really solving a lot of the problems loyalty programs see today and makes them possible even on a local or even on a regional level. In the near term, the transaction cost will get lower.

The increasing number of companies that accept cryptocurrency as payment as well as opportunity to use them at a lower cost. Japan is one of the good examples in 2017. They have yet estimated that 300,000 stores already accept Bitcoin by the end of the year 2017. There are more than 46,000 merchants worldwide that already accept Bitcoin via Coinbase, the American most popular on-ramp for the cryptocurrencies. And even the CEO of AirBnB branches said it was the most demanded feature from their community, that Bitcoin is one of the payments should be accepted.

The one area of marketing where we will see blockchain technology's impact quickly is in advertising. The fundamental changes in it through blockchain technology is much earlier. Forrester analysts say that publishers who remove mediators can increase the cost per thousand impressions from one to five dollars for themselves. Today, we have a lack of transparency between you and your intended audiences. We now have this enough visibility and how advertising is performing. Blockchain, therefore, is one of the first key-technology that can transform advertising here. In which, A customized cryptographic proof of their identity is created to know the customer interest in a related product, and they respond in time. It will probably create a marketplace, where each person will set a price for their guaranteed attention, and advertisers then will choose to pay or not to pay that individual's cost to deliver their message. If the advertisers decide they will pay, they get a response that is correct from that particular individual that we are trying to target.

The benefits of having a currency are outweighing the risks. Bitcoin, as the gold standard of cryptocurrency, started to have the first lousy reputation. But, if we begin to peel back the onion and look a little closer, every transaction is completely transparent. So governments start to see the upside of using Bitcoin and supporting the Bitcoin ecosystem. Criminals will have moved from the Bitcoin usage to other cryptocurrencies, that are providing a higher degree of anonymity. Currencies, like C-Cash or Monero, give much higher privacy protection than Bitcoin. From an economist perspective, every restriction and force inefficiency comes with a cost.

To identify industry groups and understand the appeal of blockchain technology for particular interest communities. It will help to leverage resources and best practices available from any industry to envision a new marketing approach for your targeted communities. The benefits for groups to use specific blockchain technologies are tremendous, and just starting to be explored.

The first community that formed around blockchain technology was the Bitcoin blockchain [47].

They have in common the interest of enabling trusted peer-to-peer transaction over the internet without an intermediary that binds them together. Bitcoin was an example of perfect marketing. The brand manager for the blockchain-based system, Satoshi Nakamoto, released the Bitcoin wipe out on and put the genesis block on January 3, 2009. The brand product's vision was simple: To build a community with more believer in providing cryptographic proof rather than trust [48].

The first believer in the Bitcoin opportunity was healthy. Even argued for the value that the Bitcoin network could create in the financial systems. The first that stipulated the cost of a Bitcoin could go into the millions. That attracted enthusiasts and protocols for new believers. The system grew and increased the values of the initial coins at the newly created coins, and stimulated the innovation around the newly formed technology. The second example in the blockchain environment is Ethereum. The founder of Ethereum, VitalikButerin, a Russian-Canadian programmer, created the concept and founded the Swiss-based non-profit Ethereum Foundation to drive that mission forwarded to the other industry organizations called the Hyperledger. The big company behind Hyperledger you probably have heard about is IBM. Another group that has been going to the news over the last years is called r3. Ethereum as a larger and more involved environment. They focused on refining enterprise-grade software capable of having the most complex and highly demanding applications [49]. Their key focus is to learn together and build upon the only smart contract supporting blockchain currently running in the real world, which is Ethereum. Governments themselves are starting to understand and embrace technology speed that you haven't seen not even since the internet. In 2016, the delegates from APEC improved the potential of blockchain technology, and you could see that countries like Singapore or China are already going down this route of trying to embrace it. Japan is moving very much forward as an accepted payment method as well [50]. One exciting thing we saw in 2017, governments can only influence so much. The moment individuals have entered the cryptocurrency of the blockchain world, it is much, much harder for a government. Not easy to control the on-ramp and the off-ramp of the cryptocurrency world [51–53].

11.1.1 WHAT ARE THE BENEFITS FOR A MEMBER OF THE BLOCKCHAIN COMMUNITY?

Traditional financial systems think these technologies are relatively expensive, but they have also valued use in developed countries. We will see in the emerging markets likely leapfrogging the technology adoption as we've seen with mobile technology and payment systems in China.

Milton Friedman, the famous economist, said, "Eventually, there will be a digital value system beyond the nation-state." Companies provide incremental services, the incremental value that you don't have today through the blockchain functionalities. HyperLadder.

11.1.2 BLOCKCHAIN TECHNOLOGY IN IoT

A permissioned blockchain network helps in sharing a piece of common information between the parties involved without the help of central control. On the other side, there are lots of challenges and issues being faced in establishing such a trustable network of peers in human life. The issues of blockchain management protocol, scalability, privacy, confidentiality, and interoperability of the blockchain technology. In this view, the permissioned blockchain architectures of Hyperledger: Augar and Grid Plus are the two representative decentralized application platforms. The Hyperledger is an ecosystem supporting not only blockchain protocol, the distributed ledger, and the smart contract, but it also supports the framework and tools for active engagement and collaboration of developers, businesses, and other stakeholders. The overarching goals are to promote the development of a safe, reliable, efficient, innovative, quality-driven open-source components and platform to support enterprise adoption of the blockchain technology. The Linux Foundation defines the basic framework for Hyperledger. The member organization can then design their blockchain by extending this definition. It is expected that under the umbrella of well-defined specifications, blockchain modules created by various members will be pluggable into each other's technology environment. These include standard functional modules and defined interfaces [17].

A Hyperledger blockchain application views its domain as being made up of many interacting blockchains of various capacities. The reference architecture specifies four different groups of services and the corresponding APIs for the applications to access them. List of services offered by the Heperledger framework: identity services, policy services, blockchain services, and smart contract services. *Identity services module* manages the identities of entities, participants, ledger objects, such as a smart contract. In the case of Fabric, a smart contract is called the chain code. *Policy services module* manages access control, privacy detail, consortium rules, and consensus rules. ***Blockchain services*** module manages the peer-to-peer communication protocol, the distributed ledger maintaining the global state.

The ***smart contract services*** module provide a secure and lightweight sandbox and moment for the chain code to execute. This service also provides a safe container equivalent to Ethereum virtual machine, registry, and life cycle management functions. Java Virtual Machine (JVM) that runs bytecode serves as a computational environment for the smart contract execution.

The modular design and the permissioned participation of nodes enable confidentiality, resilience, flexibility, and scalability. The fabric model consists of transactions, peers, assets, chain code, ledger, channels, identity, membership, and consensus mechanisms. Multiple business blockchains each with its channel, with its validating nodes, membership services for identity and membership management. Confidential smart contracts enable interactions between different independent business networks. The transaction could be inside a single system or cross-network. Peers are nodes that initiate transactions and maintain the state of the ledger. There are three types of peer nodes: Endorsing peers, receive and validate transactions, sign them, and return them to the creating application.

Committing peers receive the blocks created by the ordering service. Validate condition such as double spending and signature and then commit them to the ledger. Assets represent the tangible items of value that are transacted in the blockchain, for example, food supply and financial assets. Assets are represented in the program as key-value pairs in JSON on binary format. Chain-code is a smart contract that defines a set of assets and provides the functions for operating on the assets and changing the states. It also implements application-specific rules and policies.

Transactions assigned and all the access control rules set by the permission services are enforced for a transaction to be executed and included in a block for recording. Each transaction results in a set of asset pairs that are recorded on the ledger as creates updates and deletes. Since a ledger is a key-value store, it can be easily queried for later analysis and auditing Another element of the Fabric model is channels. Fabric enforces privacy and confidentiality through the channel concept. The channel defines a single permissioned network of entities with one single ledger for all its transactions and state changes. Channel provides segregated fabric for a group of objects to transact privately.

Even within the channel, the data and transaction confidentiality can be achieved by out for skating and by using cryptographic methods. Channels also provide the ability to support multi-lateral transactions among competing businesses and regulated industries through cross-chain code. Figure 11.1 describe the complete working principle of blockchain. In which client A wants to send money to other client B through peer-to-peer network (P2P). A block is created for every transaction; the P2P network helps in validating every block. This block is used to build a chain. The long-chain is valid and reliable. The structure of each block created in blockchain technology is described in Figure 11.2, along with its fields termed as a block header. The transaction made is treated as payload, which is of variable length. In Figure 11.3, the block and the interdependency of each block are clearly represented using Merkle tree.

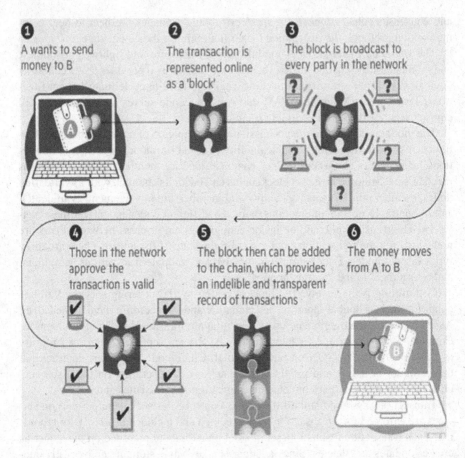

FIGURE 11.1 Working principle of blockchain. (From www.technollama.co.uk/blockchains-and-the-challenges-of-decentralization.)

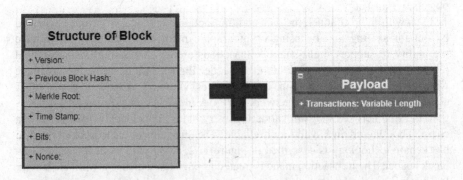

FIGURE 11.2 Creation of individual block.

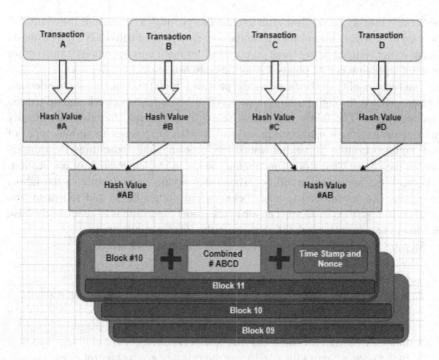

FIGURE 11.3 Merkle tree.

A Bitcoin is a public blockchain created usingP2P with the help of electronic transactions and interactions. In the traditional approach, the financial institutes are responsible for maintaining and monitoring the cash based on trust. Whereas in Bitcoin, the trust is built on the network instead of financial institutes [54]. A cryptographic proof is build using the P2P network, instead of centralized trust. The public blockchain such as Bitcoin, are permission less and unknown entities can join and leave as they wish. Entities wanting to participate in a fabric network, enroll through a trusted membership service provider (MSP). An organization manages its membership and the roles of the participating entities through an MSP. Participating entities should have a valid identity. For realizing trust, the default implementation of MSP uses an X.509 certificate as a digital identity. The peer nodes, client application, business entities, and administrators need uniform identities in the fabric. Each is assigned an identity that is an X.509 certificate. These identities determine the role of the entities and the permissions they have for accessing the resources in the blockchain network. There is a root certificate authority and intermediate certificate authorities managed by the MSP. The consensus at the high level is the agreement on the next block of the transaction to be added to the chain, and the extensive validation and verification of the order and the correctness of the transaction including double-spend and other conditions. The fabric allows for pluggable consensus model, depending on the characteristic of the channel, the minor of the next block is decided by a round-robin policy or a practical Byzantine fault tolerant or anything in between. If the channel is deemed to be highly trustworthy, a simple consensus model will do. If it is a business-to-consumer channel where there's a potential for malicious operations, even a proof of work algorithm can be plugged in as the consensus model.

The functionality of the fabric model provides these functionalities; confidentiality and privacy through building a permission blockchain network among trusted business peers. The fabric also offers segregation through the channels, identity management through membership services. It offers efficiency to various types of peer nodes operating in parallel, validating, ordering, and committing the transaction to the chain. Chaincode functionality helps in implementation of application-specific logic, and configurability through the modular architecture of the fabric. In Figure 11.4, a transaction made between the parties basha and sadiya to sent 0.7 BTC is described in detail. The job of the miner algorithm is to generate the hash needed to secure each block information in the Bitcoin application.

Steps in the development of business blockchain network are:

1. Use a personal tool to create skeleton code for the business network.
2. In the CTO file created, find the class definition for all assets, participants, and the transactions in the business network.
3. This file is written in Hyperledger Composer modeling language.
4. In the Javascript, the file is the transaction functions.
5. In the access control file (ACL), are the primary access control rules.
6. After updating the CTO JavaScript and ACL files, the entire directory is packaged using the composer tool, to package the code into a deployable business network archive.
7. Composer is used in setting the credentials, running in their server, and deploying the application.
8. The composer playground tool is executed to allow interacting with the deployed business network.

In the business application, where temperature, location, and humidity information are likely to share. A permissioned blockchain network helps in sharing

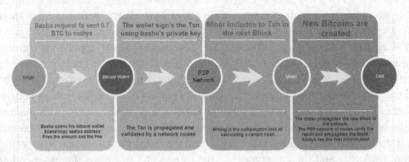

FIGURE 11.4 How Bitcoin works.

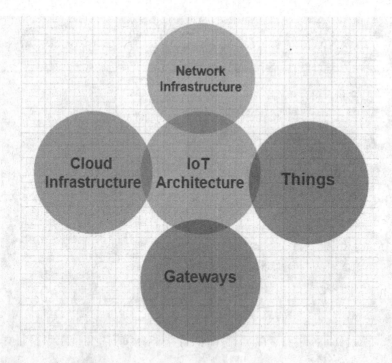

FIGURE 11.5   IoT architecture.

common information between the parties involved without the help of central control. In Figure 11.5, the overview of IoT architecture is presented.

In this chapter, we aim to provide the list of challenges that one should come across in constructing a privacy preserving model based on trust [8]. The expectation/requirements of IoT field of research is presented in Figure 11.6. Moreover, we aim to address the research questions like [9]:

RQ1: Which challenges are being raised in the context of IoT sensor data protection?

RQ2: Which are to requirements needed to design the information systems that make possible IoT sensor data protection?

RQ3: What blockchain technology to be considered?

The make the reader understand the role of data analytics and machine learning algorithms in the field of IoT is described in Figure 11.6. In which, the information flow between the basic components of IoT and cloud architecture are presented. To get the new pattern from the data collected through sensors<machine learning algorithms are used—this help to build a new response system. The research carried out in [4] is the real motivation to the present work. In which, a blockchain-based

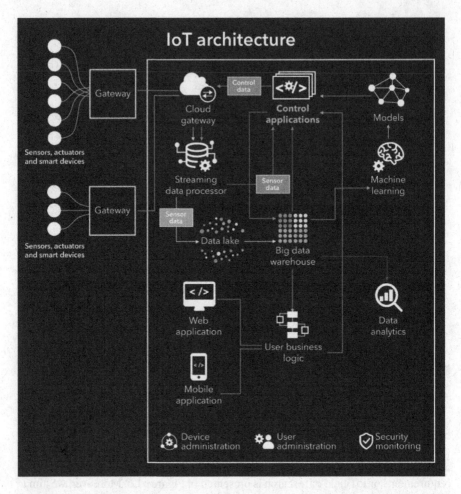

FIGURE 11.6 Basic elements of IoT architecture. (From www.scnsoft.com/blog/iot-architecture-in-a-nutshell-and-how-it-works.)

anonymous reputation system (BARS) is proposed toward preventing the distribution of forged messages and preserving the identity privacy of vehicles. A model should be able to provide the following features to provide reliable service as shown in Figure 11.7.

This chapter is organized as follows: In the Introduction section, the background knowledge required to understand the blockchain technology is discussed. In the Literature review section, the past and current research carried out in developing a privacy-preserving model using blockchain technology applied to the field of IoT is presented. In addition to that, the application areas of privacy-preserving models are discussed.

Features of a model to provide reliable service

The blockchain-enabled efficient privacy-preserving authentication mechanism.

The dependency between the public key and the real identity to be eliminated.

The certificate authority (CA) should operate the certificate issuance, and public-key revocation is to prevent the distribution of forged messages.

A law enforcement authority (LEA) should be responsible for managing BARS and storing the pairs of public keys and real identities

Need for a reputation evaluation algorithm while representing the trustworthiness of messages.

The direct historical interactions and indirect opinions about the senders to be recorded in the Blockchain as persistent evidence to evaluate the reputation score.

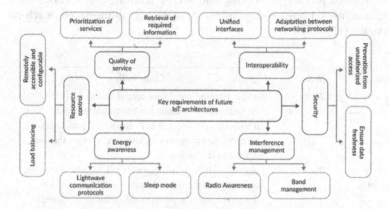

FIGURE 11.7   Requirements of IoT architecture. (From www.cypress.com/blog/corporate/future-iot-architectures-cypress-based-hands-graduate-class.)

11.2   LITERATURE REVIEW

The research made by the author in [1] is to introduce an efficient distributed privacy-preserving access control framework, in which access control models and cryptocurrency blockchain mechanisms are combined to make authorized decisions. In the proposed model, authorization tokens are considered as an access control mechanism to address cryptocurrency problems. It is making use of blockchain to ensure evaluating access policies in distributed environments, where there is no central authority and guarantee that policies will be imposed appropriately by all interacting entities. Whereas in [5], the author designed an adoptable energy management technique for smart grid network. It helps in providing the different patterns of energy consumptions without relieving the user identities.

Similarly, the author in [11] made an experiment on the relatively low level of smart home system (SHS) devices and proposed a homomorphic consortium blockchain for SHS sensitive data privacy-preserving (HCB-SDPP), based on the traditional smart home system. The features of these models are added verification services, new data structure based on homomorphic encryption. The methodology followed by the author is as follows: first, encrypt sensitive data of all gateway peers and upload them to the consortium blockchain; second, validate the security of sensitive data after homomorphic encryption processing.

A novel decentralized security model called the Lightning Network and Smart Contract (LNSC) is developed to ensure exchanges between Electrical Vehicles (EVs) and charging stations [12]. It acts as a protection safeguarding blockchain-based secure distribute/buy in plan for IoT frameworks. It empowers distributers to control information access and endorsers of get information specifically [13]. It can viably ensure information security and interests of supporters. A decentralized privacy-preserving and secure learning framework called Learning Chain uses traditional gradient descent algorithm-based machine learning systems [15].

Author Details	Year	Contribution
Ouaddah [23]	2016	An epic appropriated access control system fairness-dependent on blockchain innovation was proposed. It addresses the issues of IoT security and protection. This structure presents new sorts of exchanges that are utilized to revoke access.
Zhang [22]	2017	Proposed an E-business design for IoT, in light of the convention of the Bitcoin. Embraced disseminated self-governing enterprises as the exchange substance to manage the paid information and brilliant property.
Outchakoucht [26]	2017	Displayed the web of things worldview, and data stream in access control and security approach the executives are among the IoT gadgets. The ideas of blockchain, AI, and conveyed design are presented.
Xia [27]	2017	Proposed an access control system (ACS) for transferring sensitive information. To ensure efficient access control secure cryptographic techniques like encryption and digital signatures and permissioned blockchain are used for added security and a closely monitored system. The deployed system overcomes the traditional access control methods of passwords, firewalls, and intrusion detection systems.
Es-Samaali [29]	2017	Contributions made toward reinforcing the security of big data platforms by proposing a Blockchain-based access control framework. The concept of blockchain, mechanism and principles of the access control framework are presented with detailed explanations.

(Continued)

Author Details	Year	Contribution
Basha SM [33]	2017	Developed a model toward providing a recommendation system for PIMA diabetes victims in which machine learning algorithms are used in classifying the diabetes victims based on the feature like glucose and insulin levels.
Basha SM [34]	2017	Performed a deep analysis on feature selection methods in classifying the text reviews collected from twitters and suggested that 2 gram feature give better classification accuracy in predicting the sentiments
Novo [24]	2018	Designed of a new decentralized access control architecture for IoT using blockchain technology. Its objective is to operates as a single smart contract. The outcome of the proposed architecture is to simplify the whole process in the blockchain network and reduced the communication overhead between the nodes.
Esposito [25]	2018	Proposed an architecture toward facilitating data sharing among health information systems in which cloud computing technology is used to support real-time data sharing regardless of geographical locations and handling big data.
Banerjee [28]	2018	Made a systematic survey on security techniques that are designed for IoT using blockchain.
Wang [30]	2018	1. Proposed a structure that consolidates the decentralized stockpiling framework IPFS, the Ethereum blockchain, and characteristic-based encryption (ABE) innovation to accomplish finegrained access power over information in decentralized capacity frameworks. 2. The savvy contract is sent on the Ethereum blockchain to actualize the watchword search in the decentralized stockpiling frameworks. 3. The created framework is sent and tried under the Ubuntu Linux framework. A reproduction of the framework plan is helped out through the Ethereum authority test arrange Rinkeby, and the comparing execution and cost are evaluated.
Cruz [31]	2018	1. Proposed a role-based access control using smart contract (RBAC-SC), using blockchain technology and smart contracts. 2. Implemented a smart contract prototype of the RBAC-SC, deployed on the testnet blockchain of Ethereum.
Basha SM [38]	2018	Suggested that the identification of sarcasm before performing sentiment analysis is must do operation. That will help in improving the accuracy of sentiment classification.

(Continued)

Author Details	Year	Contribution
Basha SM [39]	2018	The aspects pair like (sentiment word, aspect word, preposition) is considered to recognize an aspect from the review. That indeed helps in performing sentiment classification at maximum accuracy.
Basha SM [35]	2018	Addressed the problems in the field of identifying the breast cancer patients. Developed an approach toward providing recommendation to the breast cancer victims.
Viriyasitavat [32]	2019	1. Provided the basic definitions and characteristics of blockchains with respect to consensus mechanisms. 2. Identified the relations of blockchain and business process characteristics. 3. Proposed an architecture and guidelines for making consensus to be more flexible and reliable in the field of business process interoperation.
Basha SM [36]	2019	Designed a methodology in adopting the huge number of reviews from the social networking platforms and deriving good weighted incentives that helps in optimizing the profitability of an organization.
Basha SM [37]	2019	Making use of open source software KNIME, performed the sentiment classification on movie reviews using decision tree machine learning algorithm.

In stock, the methodology followed in all the proposed models as per the discussion is as follows: First, investigation on authentication mechanisms. Second, to propose a security model to include registration, scheduling, authentication. Third, the security model is evaluated using real-time scenario [12].

11.3 APPLICATIONS OF BLOCKCHAIN-ENABLED PRIVACY-PRESERVING MODELS IN IOT

Let us explore different industries and functions and how blockchain technology can be implemented as part of the digital transformation today and in the long-term future. Blockchain is really for every industry. We will see it in the legal field, supply chain, government management, how energy is traded and managed and calculated. We will definitely see it in the retail space, in the food processing industry, healthcare [43,44]. Even travel hospitality, insurance business, down to education, like certificates, for example, can be approved by utilizing blockchain technology.

11.3.1 MOBILE COMMUNICATION

From 7 billion people in the world, only 4 billion own a toothbrush, but nearly 7 billion have access to a cellphone. So, mobile communication is everywhere. But only 20 % of all world population have access to the banking system currently. Now, with blockchain technology, everybody with a connection to the internet can transact

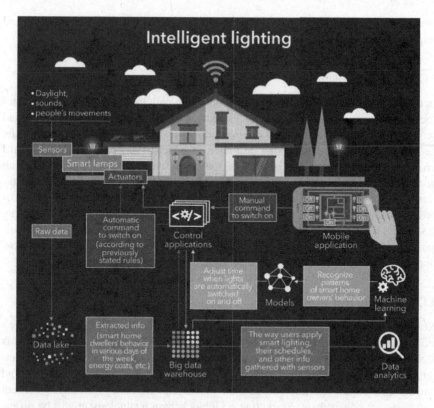

FIGURE 11.8 Example of IoT architecture. (From www.scnsoft.com/blog/iot-architecture-in-a-nutshell-and-how-it-works.)

with everybody else, allowing social scalability beyond our wildest dreams. The new economic models previously impossible before the emergence of blockchain technology are coming more and more in our daily lives. It can provide the underlying economic principles that empower a fast scaling and shared economy, demanding new banking and transaction models. Designing a smart home application using IoT is presented in Figure 11.8 in which the user has complete control over his/her home with the application installed on his/her mobile.

11.3.2 HEALTHCARE

The motivation is to make use of decentralized sharing storage system in healthcare. In the research work [18], the author aims to design a novel sharing storage system in the blockchain (Storj and Filecoin). This will help the patient to make their personal health information to propagate in the peer-to-peer blockchain network. Whereas in [2], the author applied a blockchain and enabled the method of updating electronic medical records in which distributed ledger protocol is combined with Bitcoin application. The proposed MedRec blockchain addresses the issues like slow access to medical data, system interoperability, patient agency, and improved data quality and quantity for medical research.

11.3.3 AUTOMOBILES

In the field of automobiles finding out an optimum path to charging stations of electrical vehicles is a challenging issue. To address this problem, the author in [3] had proposed a model that finds an optimum charging station without disclosing the geographic position details of the customer. In which, a blockchain is used toward providing transparency. The privacy of the proposed protocol is evaluated in an honest-but-curious adversary model. The author in [10] proposed a remote attestation security model (RASM) based on a privacy-preserving blockchain. The proposed security model is evaluated using a real vehicle to everything (V2X) scenario. It satisfied the security requirements of being decentralized, traceable, anonymous, and non-repudiatable and has excellent performance. The experimental results show that the RASM can effectively enhance the security performance of the V2X.

11.3.4 REAL-TIME BUSINESS

Now that we have a basic understanding of blockchain technology, we can envision by industry and function how everybody can be a winner with the use of this technology. Let us look how marketing for blockchain technology can teach us better community engagement. I will show you different use cases by industry and current challenges, and how we can use blockchain technology to improve client and customer engagement. Marketing with blockchain technology, it will be different. Rethink marketing and advertisement to include deeper customer engagement, rewards from customers for valuable content, loyalty and awareness programs, and creation of marketplaces.

Marketing with blockchain technology dramatically different experiences than what we have today. We will see new payment transactions modes. We will see a customer that have a cryptographic proof of her identity and the credentials. Customers can provide eligibility to shop on your site as a citizen of a certain country, or part of a certain group without telling any personal information whatsoever, through the blockchain technology. On the other hand, it allows no tracking that is imposed on customers today and provides them more privacy themselves. Advertising for blockchain innovation resembles great network showcasing. What we see currently is the genuine intensity of network to manufacture brands, in a period of hyper-associated people with feeling and energy. Eagerness for the convention draws in new devotees. The developing of biological system expands the estimation of the coin itself and invigorates its advancement. You will see an ever increasing number of brands that are brought into the world dependent on a reasonable mission, for the most part for upsetting an occupant.

A lightweight security saving ring mark plot exchanges by true clients [6]. This mark's ensures that data has not been changed, as though it was secured by a carefully designed seal that is broken if the substance were adjusted. A ring mark enables an underwriter to sign a message secretly. It uses twofold encryption of information utilizing lightweight encryption calculations (ARX figures) and open encryption plans (Diffie–Hellman key trade procedure). Making use of these both procedures together will ensure the security, protection, and secrecy of client's information through IoT gadgets.

11.3.5   Legal Sector

The legal sector has for quite some time been depending on a huge measure of information as desk work. It could profit massively from the across the board execution of savvy contracts, understandings that can be consequently initiated activities dependent on explicit conditions. For instance, the canny contract may require a driver to be modern on their vehicle installments to physically begin the vehicle. What's more, as a pattern is of most blockchain arrangements, brilliant contracts can radically lessen exchange costs and give predominant security versus customary methods.

11.3.6   Media and Entertainment

A blockchain letter could be used to secure intellectual property like music and film, making it possible to enforce usage rights and mitigate digital piracy. So, the entertainment industry is embracing it. And for them, it also creates new opportunities for sharing and monetizing their content. Makes it possible for musicians, writers, and other artists to embedded royalty payments into their digital files, their ebooks. We can reward customer's centric behavior of individuals and teams that value creation with tokens. Think verified LinkedIn data with digital signatures for experience, education, recommendations, and created in a decentralized or even outsourced team that could use blockchain-based collaborations and reward solutions to create value for their customers. The management function itself will likely be touched by blockchain technology. There might be large companies that will have no full-time employees outside of their executive-level suite. Decentralized Autonomous Organization like DAOs, those are organizations that have no formal managers at all, and just smart contracts that are executing the agreements might be another possibility. But it requires that executives excel in understanding the product, communicating the vision of the brand, the mission of the organization, the story for their community and tools and collaboration infrastructure to create and deliver value to the customers. They have to understand the whole supply chain. For example, when you buy into a shop today and buy a bag of coffee, we really don't know if it's organic free trade, even if it's written on that label.

With blockchain technology, that can be changed as well. You can be able of the provenance of a product you're buying, end to end that creates a verifiable supply chain and that can become part of the value propositions for your customers. Are companies using it? Yes. Global organizations with lots of existing business units and requirements, like Walt Disney, see the potential for blockchain technology from secure Internet of Things (IoT), to managing their logistics and Intel of finances. They have created something called Dragon-Chain, which allows enterprise programmers to utilize blockchain technology in their environment and make the technology available in open source. Juniper Research surveys for 400 top-level executives showed that 67 % said they had already invested more than 100,000 into the technology. And 91 % say they would be spending at least 100,000 in 2017 to understand what blockchain technology can do for them.

11.3.7 SECURITY

By providing an overview of different application domains of blockchain technologies in the Internet of Things. The threat models of blockchain protocols in IoT networks are classified into five main categories: identity-based attacks, manipulation-based attacks, cryptanalytic attacks, reputation-based attacks, and service-based attacks [16]. By providing a feasible solution to this attack, an author in [7] performed experiments and addressed a new location-based privacy attack in the crowdsensing system using a private blockchain to protect worker location and prevent re-identification attack. Similarly, in [14], the author proposed a blockchain-based privacy-preserving task matching scheme for crowdsourcing address the privacy-preserving task matching, identity anonymity, and reliable matching [46].

11.4   BLOCKCHAIN MANAGEMENT DEMONSTRATION

The demonstration on the creation of blockchain and its operations is made with the help of the interface built (Available: https://anders.com/blockchain/blockchain. html). In which, SHA256 hash number is generated for each block and the same is used to form collection of blocks called blockchain. It is the digital foot print of the data that it contains as shown in Figure 11.9.

Every time the same data will lead to exactly same hash value as shown in the Figure 11.10. Similarly, consider the next term called block as shown in Figure 11.11. There are four fields to address here:

1. The block field is identified a random integer value. The next block will be have the block number as the next sequence number.
2. The nonce field is used to authorized each block within a blockchain. The valid signed block will have the hash value preceding with four zeros as shown in the Figure 11.11.

Block:	#	1
Nonce:		72608
Data:		
Hash:		0000f727854b50bb95c054b39c1fe5c92e5ebcfa4bcb5dc279f56aa96a365e5a

FIGURE 11.9   Association of hash value to the message in block.

FIGURE 11.10 Representation of block in blockchain.

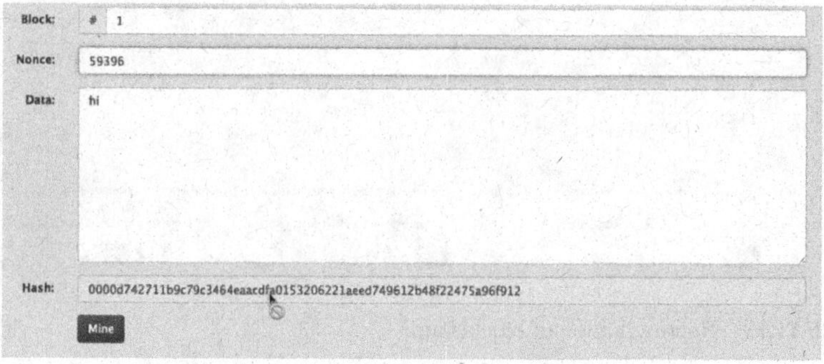

FIGURE 11.11 Identification of IN-Valid block.

The next question is how to fix the nonce value to get the valid hash value. The use of mine operation will help to fix the nonce value by changing it from one up to the value (72608). The internal operations are carried out by a mine node present in the network.

Any change in the data field will be changing the corresponding hash value. To fix it back to valid block make use of mine operation as shown in the Figure 11.12.

Such similar blocks are used to form a blockchain as shown in the Figure 11.13. For the first block the previous hash value will be all zeros indicating the first block in the chain. The second block is having previous hash value as the first block hash value. Similarly for the rest of the blocks in the blockchain.

Any change in the intermediate block, as shown in the Figure 11.14. Will make the block invalid, which can be identified by the hash value present in the current block. Hereafter all the blocks followed by the current block will be invalid. The chain has been broken from the block, in which the changes are made. To form the chain, we should mine each block after the current block to make it valid block and reform the blockchain as shown in Figure 11.15.

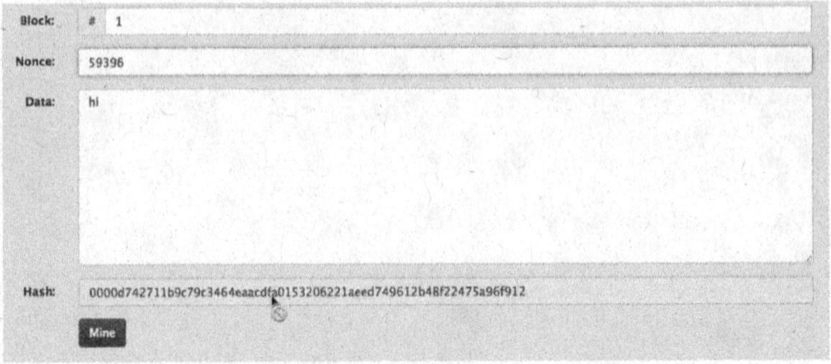

Block:	#	1
Nonce:	59396
Data:	hi
Hash:	0000d742711b9c79c3464eaacdfa0153206221aeed749612b48f22475a96f912

Mine

FIGURE 11.12   Mine operation.

FIGURE 11.13   Representation of blockchain.

FIGURE 11.14   Invalid block in blockchain.

The tokens with transactions details are represented as shown in the Figure 11.16 in which the transaction carried out by an individual is stored. In the token representation, the actual balance of account is used to validate all the transaction in the block.

In order to include the coinbase value of each transaction information carried out by each block, the peer representation is needed as shown in the

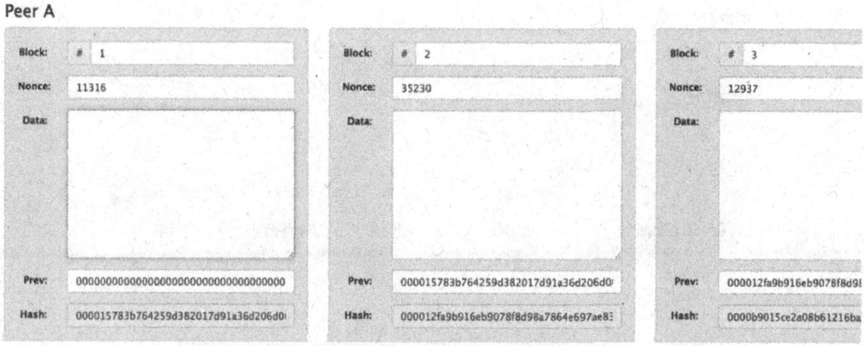

Peer A

FIGURE 11.15   Reforming the blockchain with valid blocks.

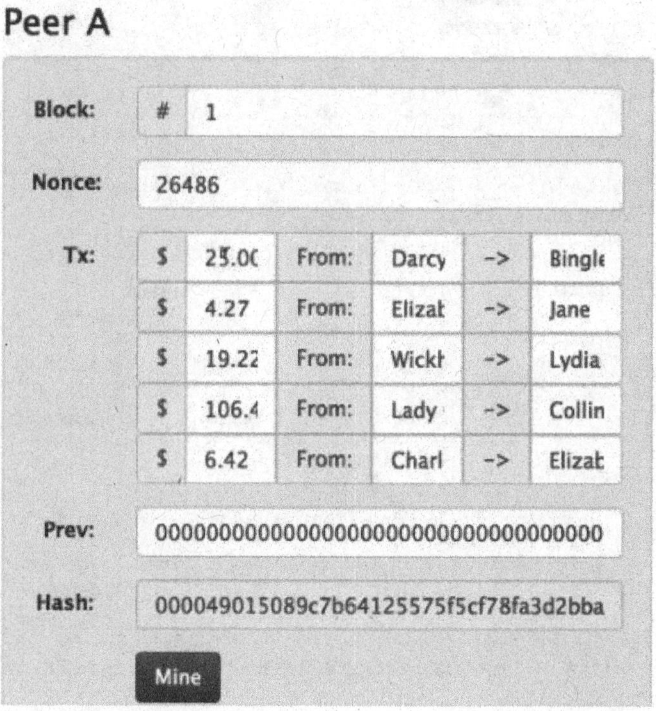

FIGURE 11.16   Block with transaction details.

Figure 11.17, in which the summation of all the transaction should not exceed the coinbase value.

Any change in the transaction made by intruder will be leading to invalidate the block and the chain will be destroyed. The longest block chain is the most trusted and valid block chain.

Peer A

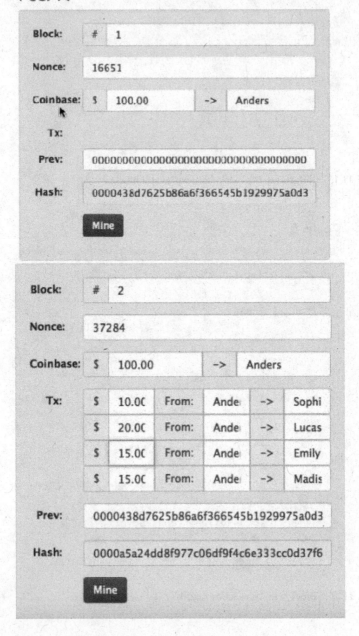

FIGURE 11.17 Use of coinbase value in block.

11.5 CONCLUSION

An understanding of the digital transformation, blockchain technology and its operations are discussed with the help of web-interface. The steps followed to understand the business requirements, embed them into blockchain technology and its overall impact on any industry is discussed. The challenges and research gaps in the field of building up privacy models using blockchain technology for (IoT) are also discussed with the help of literature review made. To the best of our knowledge, the recent research carried out in the field of developing a privacy-preserving model using blockchain technology are addressed. The limitations of the present work is that only papers from 2017 to 2019 is considered. The advantage of the present research work is to know the impact of blockchain technology on preserving privacy in the field of IoT is achieved to a certain extent. In the future, we would like to concentrate on developing a novel privacy-preserving model that applies to the field of Internet of Things.

REFERENCES

1. Ouaddah, A., Elkalam, A. A., & Ouahman, A. A. (2017). Towards a novel privacy-preserving access control model based on blockchain technology in IoT. In *Europe and MENA Cooperation Advances in Information and Communication Technologies* (pp. 523–533). Springer, Cham, Switzerland.
2. Azaria, A., Ekblaw, A., Vieira, T., & Lippman, A. (2016). Medrec: Using Blockchain for medical data access and permission management. In *2016 2nd International Conference on Open and Big Data (OBD)* (pp. 25–30). IEEE.
3. Knirsch, F., Unterweger, A., & Engel, D. (2018). Privacy-preserving blockchain-based electric vehicle charging with dynamic tariff decisions. *Computer Science-Research and Development, 33*(1–2), 71–79.
4. Lu, Z., Liu, W., Wang, Q., Qu, G., & Liu, Z. (2018). A privacy-preserving trust model based on Blockchain for vanets. *IEEE Access, 6*, 45655–45664.
5. Gai, K., Wu, Y., Zhu, L., Xu, L., & Zhang, Y. (2019). Permissioned Blockchain and Edge Computing Empowered Privacy-preserving Smart Grid Networks. *IEEE Internet of Things Journal, 6*(5), 7992–8004.
6. Dwivedi, A. D., Srivastava, G., Dhar, S., & Singh, R. (2019). A decentralized privacy-preserving healthcare blockchain for IoT. *Sensors, 19*(2), 326.
7. Yang, M., Zhu, T., Liang, K., Zhou, W., & Deng, R. H. (2019). A blockchain-based location privacy-preserving crowdsensing system. *Future Generation Computer Systems, 94*, 408–418.
8. Bahri, L., Carminati, B., & Ferrari, E. (2018). Decentralized privacy-preserving services for online social networks. *Online Social Networks and Media, 6*, 18–25.
9. Chanson, M., Bogner, A., Bilgeri, D., & Wortmann, F. (2017). Privacy preserving data certification in the Internet of Things: Leveraging blockchain technology to secure sensor data. In *Thirty-eighth International Conference on Information Systems (ICIS 2017)*, 716–726.
10. Xu, C., Liu, H., Li, P., & Wang, P. (2018). A remote attestation security model based on privacy-preserving blockchain for V2X. *IEEE Access, 6*, 67809–67818.
11. She, W., Gu, Z. H., Lyu, X. K., Liu, Q., Tian, Z., & Liu, W. (2019). Homomorphic consortium blockchain for smart home system sensitive data privacy preserving. *IEEE Access, 7*, 62058–62070.

12. Huang, X., Xu, C., Wang, P., & Liu, H. (2018). LNSC: A security model for electric vehicle and charging pile management based on blockchain ecosystem. *IEEE Access*, *6*, 13565–13574.

13. Lv, P., Wang, L., Zhu, H., Deng, W., & Gu, L. (2019). An IoT-oriented privacy-preserving publish/subscribe model over blockchains. *IEEE Access*, *7*, 41309–41314.

14. Wu, Y., Tang, S., Zhao, B., & Peng, Z. (2019). BPTM: Blockchain-based privacy-preserving task matching in crowdsourcing. *IEEE Access*, *7*, 45605–45617.

15. Chen, X., Ji, J., Luo, C., Liao, W., & Li, P. (2018). When machine learning meets blockchain: A decentralized, privacy-preserving, and secure design. In *2018 IEEE International Conference on Big Data (Big Data)* (pp. 1178–1187). IEEE.

16. Ferrag, M. A., Derdour, M., Mukherjee, M., Derhab, A., Maglaras, L., & Janicke, H. (2018). Blockchain technologies for the internet of things: Research issues and challenges. *IEEE Internet of Things Journal*, *6*(2), 2188–2204.

17. Khan, M. A., & Salah, K. (2018). IoT security: Review, blockchain solutions, and open challenges. *Future Generation Computer Systems*, *82*, 395–411.

18. Rahmadika, S., & Rhee, K. H. (2019). Toward privacy-preserving shared storage in untrusted blockchain P2P networks. *Wireless Communications and Mobile Computing*, *20*(2), 1–15.

19. www.technollama.co.uk/blockchains-and-the-challenges-of-decentralization

20. www.cypress.com/blog/corporate/future-iot-architectures-cypress-based-hands-graduate-class

21. www.scnsoft.com/blog/iot-architecture-in-a-nutshell-and-how-it-works

22. Zhang, Y., & Wen, J. (2017). The IoT electric business model: Using blockchain technology for the internet of things. *Peer-to-Peer Networking and Applications*, *10*(4), 983–994.

23. Ouaddah, A., AbouElkalam, A., & AitOuahman, A. (2016). FairAccess: A new Blockchain-based access control framework for the Internet of Things. *Security and Communication Networks*, *9*(18), 5943–5964.

24. Novo, O. (2018). Blockchain meets IoT: An architecture for scalable access management in IoT. *IEEE Internet of Things Journal*, *5*(2), 1184–1195.

25. Esposito, C., De Santis, A., Tortora, G., Chang, H., & Choo, K. K. R. (2018). Blockchain: A panacea for healthcare cloud-based data security and privacy? *IEEE Cloud Computing*, *5*(1), 31–37.

26. Outchakoucht, A., Hamza, E. S., & Leroy, J. P. (2017). Dynamic access control policy based on blockchain and machine learning for the internet of things. *International Journal of Advanced Computer Science Applications*, *8*(7), 417–424.

27. Xia, Q., Sifah, E., Smahi, A., Amofa, S., & Zhang, X. (2017). BBDS: Blockchain-based data sharing for electronic medical records in cloud environments. *Information*, *8*(2), 44.

28. Banerjee, M., Lee, J., & Choo, K. K. R. (2018). A blockchain future for Internet of Things security: A position paper. *Digital Communications and Networks*, *4*(3), 149–160.

29. Es-Samaali, H., Outchakoucht, A., & Leroy, J. P. (2017). A blockchain-based access control for big data. *International Journal of Computer Networks and Communications Security*, *5*(7), 137.

30. Wang, S., Zhang, Y., & Zhang, Y. (2018). A blockchain-based framework for data sharing with fine-grained access control in decentralized storage systems. *IEEE Access*, *6*, 38437–38450.

31. Cruz, J. P., Kaji, Y., & Yanai, N. (2018). RBAC-SC: Role-based access control using smart contract. *IEEE Access*, *6*, 12240–12251.

32. Viriyasitavat, W., & Hoonsopon, D. (2019). Blockchain characteristics and consensus in modern business processes. *Journal of Industrial Information Integration*, *13*, 32–39.

33. Basha, S. M. (2017). A soft computing approach to provide recommendation on PIMA diabetes. *International Journal of Advanced Science and Technology, 105*(1), 19–32.
34. Basha, S. M., & Rajput, D. S. (2017). Evaluating the impact of feature selection on overall performance of sentiment analysis. In *Proceedings of the 2017 International Conference on Information Technology* (pp. 96–102). ACM.
35. Basha, S. M., Rajput, D. S., Iyengar, N., & Caytiles, D. R. (2018). A novel approach to perform analysis and prediction on breast cancer dataset using R. *International Journal of Grid and Distributed Computing, 11*(2), 41–54.
36. Basha, S. M., & Rajput, D. S. (2019). A roadmap towards implementing parallel aspect level sentiment analysis. *Multimedia Tools and Applications, 78*(20), 29463–29492.
37. Basha, S. M., Rajput, D. S., Thabitha, T. P., Srikanth, P., & Kumar, C. P. (2019). Classification of sentiments from movie reviews using KNIME. In *Proceedings of the 2nd International Conference on Data Engineering and Communication Technology* (pp. 633–639). Springer, Singapore.
38. Basha, S. M., & Rajput, D. S. (2018). Parsing based sarcasm detection from literal language in tweets. *Recent Patents on Computer Science, 11*(1), 62–69.
39. Basha, S. M., & Rajput, D. S. (2018). A supervised aspect level sentiment model to predict overall sentiment on tweeter documents. *International Journal of Metadata, Semantics and Ontologies, 13*(1), 33–41.
40. Balakrishnan, S., Parvathy Nathan, S., & Manoshree, J. (2019). Cryptocurrency and cryptography. *CSI Communications Magazine, 43*(6), 19–21.
41. Balakrishnan, S., & Janet, J. (2019). Blockchain technology: Basics, architecture, use cases and platforms. *CSI Communications Magazine, 43*(5), 10–14.
42. Nagasubramanian, G., Sakthivel, R. K., Patan, R., Gandomi, A. H., Sankayya, M., & Balusamy, B. (2018). Securing e-health records using keyless signature infrastructure blockchain technology in the cloud. *Neural Computing and Applications, 32*(3), 639–647.
43. Kumar, S. R., Gayathri, N., Muthuramalingam, S., Balamurugan, B., Ramesh, C., & Nallakaruppan, M. K. (2019). Medical big data mining and processing in e-healthcare. In *Internet of Things in Biomedical Engineering* (pp. 323–339). Academic Press, India.
44. Dhingra, P., Gayathri, N., Kumar, S. R., Singanamalla, V., Ramesh, C., & Balamurugan, B. (2020). Internet of Things–based pharmaceutics data analysis. In *Emergence of Pharmaceutical Industry Growth with Industrial IoT Approach* (pp. 85–131). Academic Press, India.
45. Kumar, S. R., & Gayathri, N. (2016). Trust based data transmission mechanism in MANET using sOLSR. In *Annual Convention of the Computer Society of India* (pp. 169–180). Springer, Singapore.
46. Muthuramalingam, S., Bharathi, A., Gayathri, N., Sathiyaraj, R., & Balamurugan, B. (2019). IoT based intelligent transportation system (IoT-ITS) for global perspective: A case study. In *Internet of Things and Big Data Analytics for Smart Generation* (pp. 279–300). Springer, Cham, Switzerland.
47. Rana, T., Shankar, A., Sultan, M. K., Patan, R., & Balusamy, B. (2019). An intelligent approach for UAV and drone privacy security using blockchain methodology. In *2019 9th International Conference on Cloud Computing, Data Science & Engineering (Confluence)* (pp. 162–167). IEEE.
48. Khari, M., Garg, A. K., Gandomi, A. H., Gupta, R., Patan, R., & Balusamy, B. (2019). Securing data in Internet of Things (IoT) using cryptography and steganography techniques. *IEEE Transactions on Systems, Man, and Cybernetics: Systems, 50*(1), 73–80.

49. Karthikeyan, S., Patan, R., & Balamurugan, B. (2019). Enhancement of security in the Internet of Things (IoT) by using X. 509 authentication mechanism. In *Recent Trends in Communication, Computing, and Electronics* (pp. 217–225). Springer, Singapore.

50. Shankar, A., Jaisankar, N., Khan, M. S., Patan, R., & Balamurugan, B. (2018). Hybrid model for security-aware cluster head selection in wireless sensor networks. *IET Wireless Sensor Systems*, *9*(2), 68–76.

51. Karthikeyan, S., Rizwan, P., & Balamurugan, B. (2018). Taxonomy of security attacks in DNA computing. In *Advances of DNA Computing in Cryptography* (pp. 118–135). Chapman and Hall/CRC, India.

52. Namasudra, S., Devi, D., Choudhary, S., Patan, R., & Kallam, S. (2018). Security, privacy, trust, and anonymity. *Advances of DNA Computing in Cryptography*, *25*(2), 138–150.

53. Rahman, M. A., Ali, J., Kabir, M. N., & Azad, S. (2017). A performance investigation on IoT enabled intra-vehicular wireless sensor networks. *International Journal of Automotive and Mechanical Engineering*, *14*(1), 3970–3984.

54. Bhuiyan, M. Z. A., Wangy, G., Tianz, W., Rahman, M. A., & Wu, J. (2017). Content-centric event-insensitive big data reduction in Internet of Things. *IEEE Globecom*.

12 Convergence of IoT, AI, and Blockchain
Technology Pathway to Swachh Bharat

P. Hamsagayathri, K. Rajakumari,
P. Ramya, and K. Shoukath Ali

CONTENTS

12.1 INTRODUCTION

Waste is produced at any place wherever people are living, and it is the part and parcel of the everyday routine lifecycle. "As per World Bank's survey report [1], in 2012, the worldwide Municipal Solid Waste levels will be around 1.3 billion tons per year and it is increasing every year and become a serious concern of the society." Due to the scarcity of agricultural resources, people are moving toward urban areas, and it will cross 70% in the developing countries in the forthcoming years. With the increased modernization and change in the human life style, generation of waste is increasing, and it has turned into a high challenge for developing countries [2]. Urban waste management associations in various cities face the test to give proficient and persuasive framework to collect, dispose, sort and reuse the waste to keep up healthier and friendliness environment. In developing countries, waste collection and disposal of waste management systems are affected by improper waste collection in litter bins, insufficient collection schedule, and route planning and inadequate resources [3]. Due to scarcity of equipped waste containers, people started dumping the waste in open areas and roadsides [4,5].

With respect to recycling, social and economic factors forms standard factors to build up strong and efficient recycling system. In addition to that, technical and financial factors also influence the efficiency of the waste management system. Better innovation and better methods for dealing with waste paves the pathway to a systematic approach. Advancement in waste management is essential to provide sustainable waste services to society which in turn build up an eco-friendly environment. Emerging technologies can be used for efficient waste management and helps in recognizing the industry stakeholders for recycling. It also forecasts and informs about the type, quantity and location of waste to stakeholders and helps them to plan the process well in advance. Moreover, waste-related information needs to be stored in the cloud, where it can be easily accessed by stakeholders to analyze the data efficiently.

Another important issue in waste management is to reduce the waste collection time, schedule the routes and balance the workload among vehicles. The huge volume of waste that is generated throughout the world has driven governments in different nations to make severe strategies, to guarantee effective waste collection and disposal. In addition, the Indian government also reframed the laws toward waste management. These regulations set up eco-friendly environments with proper disposal of municipal solid waste.

There are numerous categories, each with different classifications of waste materials, like clinical to nuclear, biodegradable to nonbiodegradable and common household to industrial toxic waste. While developed countries are able to manage and treat these waste materials of different categories, developing countries like India are still struggling with the collection and proper disposal of common household waste materials. Disorganized management and dumping of waste is a noticeable cause for ruining the environment in the major cities of these developing countries. High-volume of industrial solid wastes includes the different chemicals, petroleum products, metals, leather, and so on. The waste from auto equipment shops, construction firms, cleaners and pesticide applicators are also harmful to the environment. Improper management of these wastes results in environmental pollution, which may cause substantial damage to public health and welfare. Most of the domestic waste consists of food and kitchen waste, green waste paper that is biodegradable, and there are also nonbiodegradable bottles and plastics.

The pictorial representation of smart waste management system is depicted in Figure 12.1. Due to many socioeconomic factors, implementing existing smart solutions for waste management systems is a great challenge in developing countries. Waste that are thrown improperly lead to unhealthy. Therefore, both domestic and industrial wastes need to be packed, dumped, collected, transported, and recycled properly and hence waste becomes a wealth of the country. Using promising technologies, real time data from waste bins are sent to the cloud and it is stored. Separate waste bins for different waste categories such as organic, plastic bottles, and metal are provided, and sensors are embedded in these bins to sense the waste quantity and update its status of waste level to the cloud. This helps stakeholders to efficiently and conveniently to handle waste. This also helps the waste management to adopt suitable route plan for waste collection in smart metropolitan cities.

In this chapter, design of an effective smart waste management system based on IoT, AI, and blockchain technologies in perspective of developing countries for smart decision making is proposed. The system adhere efficient waste collection, disposal, sorting and recycling of waste. Blockchain offers smart contracts for user

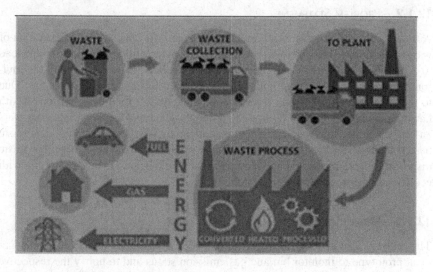

FIGURE 12.1   Smart waste management.

transactions, where these contracts are self-executing PC codes that take indicated activities when certain conditions are met in reality. Waste management using IoT, AI, and blockchain technologies bring more coordination among public and municipality boards [6–9]. It will facilitate the government to control the waste recycling process. These technologies put hand together to reduce the imbalance between different sectors and that leads to increased transparency throughout the process. Hence, the efficient waste management system is essential for the hygienic society and for a better world as a whole.

12.1.1 Motivation

According to World Bank's statistics from 2012, it forecast that 4.3 billion urban inhabitants will create about 1.42 kg/capital/day of metropolitan strong waste by 2025 [1]. The waste administration expenses are anticipated to increment every year. The development and industrialization of the country increases the waste generation appropriately. At present, around half of complete populace of world is dwelling in urban areas. Hence, waste management is going to be important aspect of government. Moreover, in many developing nations waste is spread over streets because of inappropriate techniques for gathering and dumping therefore polluting the earth. Due to socio-economic drawbacks there exist problems in primary task of waste management like proper disposal, collection, sorting, recycling, etc. These existing problems have motivated us to use the emerging technologies to lend their hands to achieve the smart city using smart management systems. Such systems are designed using Internet of Things (IoT), artificial intelligence (AI), and blockchain technologies. The IoT interconnects objects and people in the environment through the internet, which places its footstep toward economy digitization. Further, it creates new business models with wide implementation scope.

12.1.2 Problem Statement

A traditional waste management system has several drawbacks where generation of waste is unpredictable. There are some litter bins which fill up much faster and lead to overflowing before the next scheduled time for collection, which pollutes the environment. Moreover, frequent collection is required during special occasions. Due to the accumulation of waste, litter bins produce toxic which cause serious health hazard to human habitation. This study proposes a technical solution that empowers an efficient waste management model that automatically detects and collects waste based on bin fill and gas emission status. The proposed model helps in proper waste disposal, collection, sorting, and recycling that would help to hype a smart city with eco-friendly waste management system.

12.1.3 Objective

The primary objective of the chapter is to design a high-level smart-waste management prototype to monitor bin and gas emission status and to notify the respective personnel to approve the truck request for waste collection. It also automatically detects and sorts the waste for recycling. Moreover, it rewards and encourages the people for proper waste disposal and provides biomanure for resident farms to achieve ecofriendly environment.

12.1.4 ORGANIZATION OF THE CHAPTER

The rest of the chapter is organized as following. Section 12.2 describes the related work. Technological background is provided in Section 12.3. Section 12.4 presents the high-level architecture of the waste management system. Challenges are highlighted in Section 12.5. Finally, conclusion and future scope are detailed in Section 12.6.

12.2 REVIEW ON STATE OF THE ART

Waste management is a core task that occurs every day in different domains, and it becomes an important requirement for ecologically sustainable development in many countries. Moderate amounts of waste are generated in an environment inhabited by human beings [10]. A well-organized management of waste is a serious issue in today's world [11]. The improper solid waste management affects the urban settlements, mostly cities and big municipalities, which are catastrophic in nature [12].

Waste bins are used is the common method for collecting of solid waste disposal [13]. Both economic and commercial waste products are collected and disposed at particular functional point [14]. The common waste collection process demands that waste management personnel has to pass by and collect waste from all the different waste bins. The waste management personnel have to present themselves in person at all the waste collection points without the knowledge of the bin status. In this case, two possibilities present—either there is no waste for collection, or the bin has overflow. This is a complex and time-consuming process. With rise in waste disposal expenses and high perceivability of waste collection process are enforcing the residents to look for efficient system for solid waste collection and disposal [14,15]. Garbage bins are cleaned on time to maintain a hygienic environment [16].

With the dawn of new technologies, the research and development of smart cities is in progress and it is still is underway. The most significant related work done using innovative techniques were presented here.

Author [17] provided the survey on different waste management system of 22 countries across the world. The authors concluded that stakeholders and socio-economic aspects are the key factors that affect the each and every step in waste management system such as "collection, segregation, transportation and disposal." Authors also highlighted the significance of an effective and more intelligent method for analyzing the waste by organizations for its recycling. A novel waste management method [18] was presented by incorporating the Social Network Analysis with Stakeholder. The result of the study recommends about concerns of stakeholders about the communication in the waste management and they look for development in such manner. Besides, stakeholders need to be involved in designing the framework, which helps them to identify the co-stakeholders in particular domain. The service recipients should be made legitimately part of the system, so as to accomplish efficient waste management.

Author [19] expressed his mind-boggling thoughts where circumstance of quickly developing population, increment in movement, instable circumstances in different nations, unavoidable change in the atmosphere, vitality and asset constraints, and so forth represent a test in tending to assorted interests, qualities, and targets, inalienable among stakeholders. In this way, an increasingly proficient and compelling instrument is required, with the end goal that the stakeholders know about what is significant to them and in what measure. Stakeholders can from that point get ready and viably handle the waste.

The waste management system was examined in New York [20]. They expressed that public well being was a key driver of waste practices in the United States. But currently, ecological concerns have moved it down. They also demonstrated the significance of waste management system. Authors in [21] also focused the same viewpoint in their discussion about U.S.'s waste management practice. Authors [22] emphasized the significance of wastecollection time, due to waste decomposition greenhouse gases are discharged in to the air, which cause environmental change across the world. Emission of greenhouse gases can be controlled only with appropriate notification and collection of waste.

A detailed review was provided on the solid waste recycling approach in Malaysia [23]. They stated that still Malaysia highly depends on the land-filling process for waste disposal. This has brought about space constraint, medical problems [24–26], and environmental issues. Probably the most ideal approaches to handle the reusing issue in Malaysia and other such countries is to have a legitimate notice and information accessibility, with the goal that the sort and amount of Materials for recycling is known and partners are engaged with the procedure in a successful manner. Authors [27] presented an investigation on portraying metropolitan Solid waste in Kuwait. They noticed that the "everyday normal of wastage is 1.01 kg/individual". "The majority of the waste is of natural issue, involving 44.4%. Rest is made out of 11.2% film and 8.6% of layered strands as the imperative sorts of waste." So as to have a total waste management methodology, it is critical to have a keen method for informing the amount of each sort of waste and include the stakeholders viable. Similarly, authors [28] also concluded the same by exploring European nations waste management systems. In their work [23] specified that one of the significant applications of IoT in urban areas is the food industry. It is essential to analyze the food business, and it is conceivable by monitoring the natural waste. Provenance of waste additionally assumes a significant job in overseeing food industry and other related procedures.

There exist different possible solutions that were intended to provide effective waste collection techniques with efficient and optimized route planning for logistic purpose. A "geographical information system (GIS) transportation model for solid waste accumulation" explains plans for waste stockpiling, gathering and transfer has been proposed in [51] for Indian cities. In [29] authors proposed enhanced routing and energy efficient scheduling waste collection model for the Eastern Finland. Authors proposed [30] a truck-based scheduling waste collection model for Brazil. The purpose of the research is to investigate and build up optimal scheduler trucks for well defined waste collection routes. Different frameworks models were portrayed in [31–59].

Authors investigated waste management practices of developing nations from 2005 to 2011 and a comprehensive survey displayed in [60]. They also analyzed difficulties that exist in the waste gathering circle of developing nations. The research was focused on purpose of the stakeholders activities and assessment of factors that characterizing their job in waste collection process. The models in the overview were tried on genuine information. In [31] a approaches of municipal waste collection in budding nations were displayed. The exploration exhibited data from 1960s to 2013. The challenging key points and design complexities were also discussed. Finally, they concluded that significance of a solid waste management system is incredible in developing nations. The major issue is waste accumulation where emerging methodologies were propelled and can be utilized for planning and steering the business. Information about the containers was not considered for waste collection. Most of the literature reviews proposed a model that utilized IoT and AI innovation for smart cities. In addition, enabling the stakeholders, waste management personnel, and public in single framework was considered. All these issues permit researchers to build up advanced framework by encouraging the utilization of IoT information, dynamic directing models, and blockchain technologies to support of differing sorts of stakeholders and society.

Radio frequency identification technology (RFID) and wireless sensor networks (WSN) were implemented to minimize the uncontrolled waste disposal [51]. Some authors have used RFID technology to identify and track selective collection by storing owner data and information about their bins. They proposed a method to improve the quality of selective waste collection by tracking the waste stream of a city, where each waste is detected from information stored on an RFID tag (associated with waste), and during the waste processing step RFID tags are read to provide some relevant information. The household waste volume was estimated based on image analysis of the contents of the open container lid from RFID tags, such as, the label would be used to associate each bin to the address of the house that owns it [30].

An approach was proposed to classify garbage in to six different recycling categories such as "metal, paper, glass, plastic, trash, and cardboard" by using support vector machines (SVM), neural network (NN) along with scale-invariant feature transform (SIFT) features were used [41]. Their approaches achieved an "accuracy rate of 63% and 22% for the trained SVM and the CNN," respectively. Our work is similar to the work of Yang the images are trained with others NN techniques [38–44], such as, support vector machine (SVM), pre-trained VGG-16 (VGG16), AlexNet, K-Nearest Neighbor (KNN) and, random forest (RF). Waste classification method was developed to categorize the different wastes in to three categories (e.g., landfill, recycling and, paper) of an image of a jumbled waste by using faster region-based convolutional neural networks (Faster R-CNN) technique to obtain region of interest to classify objects [42].

Neural network is considered as one of the most familiar deep-learning algorithms for various applications in image processing [31–33, 60,61] and hence, convolution neural network (CNN) was used to perform waste classification.

Support vectors and neural network classification models [29,41–43] were used to classify municipal waste into different categories and they achieved an accuracy rate of 63% and 23%, respectively. GoogLeNet-based vision application was developed to confine and classify residential wastes [45]. They experimented to classify different waste types and accomplished accuracy rate ranging from 63% to 77%

respectively [46]. Google's TensorFlow and digital camera [47] was used to categorize waste typeautomatically for composting and recycling process. However, there
does not exist experimental results.

The idea of blockchain was presented in [49] and its working model along the
details of security and privacy, have been referenced in [52]. Different uses of blockchain are depicted in [50–54]. The working principles of Ethereum blockchain have
been depicted in [10–13]. Few uses of blockchain and smart contracts have been referenced in [14–16]. From various literature survey, a smart waste management system using emerging technologies like IoT, AI, and blockchain has not been proposed.

12.3 BACKGROUND

12.3.1 IoT

Internet of Things (IoT) has emerged from machine-to-machine (M2M) communication, where it interconnects different objects through a network without human intervention. M2M associates different wireless devices to the cloud for collecting and
managing data. M2M is further developed and it use sensor networks in smart devices
to connect with people and other devices. Many industries utilize IoT to enhance
their business productively, where it is able to understand customer requirements, and
appropriate decisions are made accordingly to improve the business value.

The primary objective behind the IoT is to have devices that self-report in real
time, and conveying information to the outside world more quickly without human
intervention. The Internet of Things had transformed a wide range of fields. In medicine, IoT helps medical professionals to monitor patients, and it also let them to
monitor remotely. IoT data can be further evaluated to help practitioners, so that they
can adjust treatments to improve patient outcomes.

IoT has propelled much faster, however, there are certain potential risks faced by
companies that have yet to be addressed and it limits its proceedings to expand the
range of internet-connected devices.

IoT is extremely impressive, in reality it does not serve much without a good artificial intelligence system to take advantage of it. Both IoT and AI need to develop
at the same rate to function perfectly. It also explores new data analysis software
devices that operates smarter for a safe and effective Internet of Things to come true.

The integration of artificial intelligence into the functions of the IoT is fast becoming a condition for digital ecosystems success based on today's IoT. Therefore, companies have to move quickly to identify how they will gain the value of combining
AI and the IoT or face up to its updating in the coming years. The only way to keep
pace with this data generated by the IoT and obtain the hidden knowledge it contains
is to use artificial intelligence as the ultimate catalyst for IoT.

12.3.2 AI

Artificial intelligence (AI) is an emerging technology that builds the machine to
learn from experience and adjusts its inputs to perform human specific tasks accordingly. The computers are trained and process large volume of data to simulate human

intelligence in them. Nowadays, AI has become more popular because it automates repetitive learning and analyzes the data deeper and discovers the patterns with incredible accuracy. Hence it is used across the different industry sectors. With the advent of AI, process of garbage sorting and recycling is automated. In a smart waste management system, trash bins are embedded with IoT sensors to measure waste levels, and status of the trash bins are send to server through Internet for further processed and optimized for waste disposal. Using different AI algorithms information about trash rate can be studied and that helps to predict the various factors like temperature.

Presently, steel, iron, aluminum, food, and paper waste can be separated easily in municipal solid waste using automated algorithms. However, removal of glass and plastic containers from solid waste is manually done by blue-collar workers. By using artificial intelligence techniques, a machine can be built to automate the process of efficient removal of plastic and glass containers from solid waste.

In addition AI techniques were followed to reduce the waste collection time of vehicles and to balance the workload and schedule the routes among vehicles. With this aided technology, multiple vehicles cooperate among one another and collect the solid waste efficiently. The wastes in the streets may be either dumped on single side or double sides of the road. There also exist roads/streets where it required collecting waste once or twice per day based on the waste disposal. By considering various environmental and traffic conditions, an efficient routing protocol can also be built based on AI technology. AI-based robots are activated at recycling industries to segregate the waste for further process. AI machines are greatly relying on historical data that helps them to refine the system efficiency. Steps involved in AI-based waste recycling are depicted in below process flow Figure 12.2.

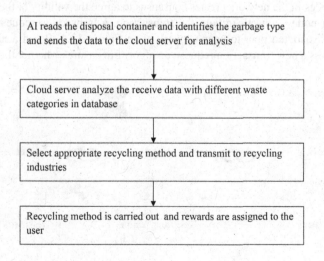

FIGURE 12.2   AI-based waste recycling.

12.3.3 BLOCKCHAIN

Blockchain is a decentralized, distributed, transparent digital ledger used for secure transfer of money, property, contracts, and so on. It uses decentralized consensus and encryption techniques to maintain the secured network. The most popular application of the blockchain technology is Bitcoin, which is used to exchange the product and services. However, blockchain and Bitcoin are different: Bitcoin is a digital token, whereas blockchain is a publicly accessible ledger used to store the transactions. Transactions on blockchain exchanges digital currency and all verified transactions are stored in the form of blocks.

Hash techniques are used to link the blocks together to form a chain of blocks as shown in Figure 12.3. Each block holds data, hash value of current and previous blocks. Any change in the block always results in generation of new hash value for that particular block. As each block in blockchain is linked with hash, it serves as bullet proof against the attackers. Moreover, a copy of the blockchain is maintained at each node in the network, which helps them to initiate and verify different transactions. Nodes in the network use wallet software to create the cryptographic keys with one private and public key, respectively. The public key of a node is shared to the network where it acts as a unique identifier for transactions. A node that wants to transact the money creates new transaction with its own and receiving node's public key. Due to the advancement in system performance, intruder can easily tamper the data and reconstruct the hash key in corresponding blocks at a fast pace to make valid blockchain. Proof-of-work (PoW) concepts are used to overcome such issues in blockchain, PoW involves high computation methods, which slows down the creation of new blocks in the chain. Suppose, if an intruder tampers the data in a particular block, PoW need to be recalculated for all the blocks. Hence, it sets up a concrete wall to secure the data in blockchain. In addition, blocks in blockchain constitute distributed peer-to-peer network to secure themselves. The change in a block is broadcast across the network, and each node authorizes the block change so as to add the block in the blockchain. Hence, all these nodes in the network creates consensus to agree the validity of the blocks.

When a node wants to initiate the transaction, a transaction request is broadcasted to distributed peer-to-peer network as in Figure 12.4. When a neighboring node receives a new transaction request, an algorithm ensures whether the block was

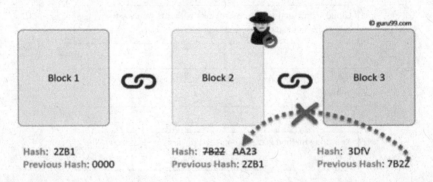

FIGURE 12.3 Hashing in blockchain.

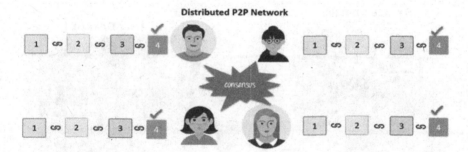

FIGURE 12.4 Consensus.

actually signed by the creator node and user status to verify transaction integrity. Finally, the new block is added to the blockchain after computing the hash. It is very difficult to calculate the hash value lower than the threshold value. The static block contents in the blockchain always ends up with same hash and variable value. Nounce are added to the block contents to find the new hash value below the threshold value. Miners are used in the network to identify the suitable nonce to add in new block and this process of searching nonce and miners is called mining, where it requires high-speed processing units. The new blocks mined by miners are relayed over the distributed network, and it will be appended to blockchain only when the majority of peers agree on the consensus of the new block.

Bitcoin is the first footstep to introduce blockchain to the world. It allows for the financial exchanges across the borders and totally replace the third party for security and exchange overhead cost. In spite of the fact that bitcoin is effective and advancements in blockchain have exhibited their value in both financial and non-financial frameworks. The use of smart contracts over blockchain shows the extent of the effect of blockchain. The decentralized autonomous organization (DAO) is a design where it can exist on Ethereum blockchain and whose target is shown based on a collection of smart contracts. It keeps running alone without human intercession. Any changes made to the association must pass through all individuals or in all likelihood it won't happen. A DAO can connect with different DAOs and execute smart contracts in them. This ascent new plan of action with wide execution scope. Even government bodies and various organizations can envision an application to run over blockchain using the idea of DAO.

12.4 DESIGN AND DISCUSSION

12.4.1 System Architecture

The TAG is managed by Ethereum blockchain. The smart contracts execution is the real motivation by which Ethereum drives out DAO. Three distinct languages such as Solidity, Serpent, and LLL are used to formulate smart contracts. Once coded, it is compiled and deployed in Ethereum blockchain. Solidity is a Java-based language that was created and upheld by the Ethereum people group. Serpent and LLL are Python-like execution and Lisp-like usage, respectively. Mostly, Solidity 0.4.8

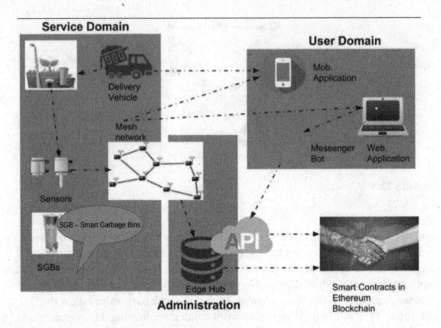

FIGURE 12.5 Waste management system architecture.

is used to create smart contracts for TAG. The applications on Ethereum are manufactured using Truffle and Embark systems. TAG collaborates with three smart contracts and it is deployed using Truffle v2.1.2. The overall architecture of a smart waste management system is shown in Figure 12.5.

12.4.1.1 Bank

The bank DAO is the portrayal of any certifiable association over blockchain. One of its significant aspects is the functional code of the blockchain cannot be altered after deployment. Hence it is difficult for the intruder to tamper the data. Individuals can change the principles or goals utilizing the voting mechanism. Such bank serves as smart contracts in real-world blockchain applications. The bank has the registries that hold addresses and balances of different parities. Each address has a unique account number to access the users. The addresses can be a record number of clients or different DAOs. This bank can be comprehended as a central bank that regulates the economy of DAOs. However, common currency "PercCoin" is used across different DAOs. Such currency is created using two techniques, and it can be easily implemented by accessing the codes from Github.

12.4.1.2 External Transfer

Trade rates can be characterized for various monetary standards used in the bank. Whenever users transfer a certain amount to bank, sender account information are updated in the registry and with appropriate measure of PercCoin with a conversion scale 1 Wei = 1 PercCoin is updated to the existing account balance of the user.

12.4.1.3 Minting

Ethereum creates the Ether tokens to create the blockchain where it can be transferred between the accounts to compensate the processing time of mining nodes. Mining the Ether requires more computational resources and it is highly expensive. Moreover, it also increase the network complexity. The bank can mint a comparable measure of equivalent currency to add it to the client's record.

12.4.1.4 Community DAO

Community DAO can be comprehended as an association running to manage the waste in efficient way. All the proposals of members in community DAO is elected based on voting. All the voting rules are set during initialization and voting rules can also be updated based on new proposal request. Ethereum gives distinct layouts for different organization. A Congress format given in the Ethereum site is utilized here. The owning address can include and evacuate members based on the voting capabilities.

12.4.1.5 minimumQuorum

Quorum is one of the primary significant aspects toward normal adoption of blockchain among monetary ventures. Moreover, it is enterprise-centered, permissioned blockchain framework intended for financial transactions. These permissioned blockchains creates trust and operates between approved nodes [62]. minimum quorum represents the least number of members required to approve the new proposal from the member of Ethereum.

12.4.1.6 debatingPeriodInMinutes

It is the maximum time limit set for users to vote for particular proposal. Suppose, if *debatingPeriodInMinutes* is set as 5, all the members in communityDAO need to vote within 5 minutes.

12.4.1.7 majorityMargin

A new proposal requires sum of 50% of the vote and majority margin number of members to pass through communityDAO [63]. Suppose if there are 10 members in the community and if the *majorityMargin* is set to 1, then the total number of votes required to pass through the proposal is 6. However, for absolute majority, the margin is set to −1 or it simply left to 0.

12.4.1.8 SGB Factory

SGB Factory holds a record for SGB makers. It tends to be comprehended as a commercial center point and communityDAO can also support by raising funds. The individuals can choose their own investment amount for SGB and pass a proposal to transact fund amount to this SGB Factory. Once SGB producers receive funds, they transfer the same to SGB, which is one of the key features of blockchain. The blockchain transactions are used for exchanging goods and services.

12.4.1.9   SWM Server

Despite the fact that, blockchain has intense potential, were its advancement need to be acknowledged. The implementation of blockchain is highly complex where it requires more processing time and hence it undermined for high-speed applications. Due to this drawback, design of TAG is made hybrid, where a centralized server, smart waste management server or SWM handles parts that need quick read compose and blockchain innovation to deal with the user transactions. The SWM has the parts that are discussed in the following sections.

12.4.1.9.1   Telegram Bot

TelegramBot is a bot application that collaborates with both user and bots by sending command requests. User can connect with bot to determine balance or to produce QR-code that is expected to open the SGB. In the event that the clients have enough balance in their record, this bot creates a QR code and requests clients area. When the area is given, it is considered as the focus of the circle and the nearest SGB is identified and returned back to the client. The geoJSON and GIS queries in Firebase database are used to fetch geographical data.

Users have the option to go to the SGB to search the QR-code that had already been generated. When a generated QR-code is not reused, then the Telegram bot automatically opens up the new SGB connectivity.

12.4.1.9.2   API

A smart waste management application offers different API to communicate with user and bots with various data endpoints. The fundamental APIs used to retrieve the waste collection details are /api/info/:sgb_id and /api/filled_bins/:percent/:radius/:lat/:lon, where it returns current location, waste amount, rate for given SGB id.

12.4.1.9.3   SGB Simulation

The simulation environment is set up using SGB simulator and QR code scanner. MQTT subscribe and publish queuing technique is used to connect to the server. The simulator examines the QR code produced by the Telegram Bot in client's Telegram application and speaks with the server utilizing MQTT convention. Lids of SGB are opened with decoded QR code sent by SGB.

12.4.1.9.4   FireBase

Firebase is a cloud-based real-time database and data from each user is constructed in JSON format and stored in it. Rather than common HTTP demands, Firebase real-time database also provides data synchronization for each time data changes, and it will be updated in inside milliseconds in the devices. It supports both online and offline mode, in case of latter, when network is restored, the customer gadgets are synchronized with the present server state.

The SGB information like user account, waste-related information, geographical information, etc., is stored in the Firebase database. The geoJSON and GIS queries in the Firebase database are used to fetch geographical data. data related to geographical details can be easily fetched with location value and its radius. Whenever the user selects the option to throw the waste, all these geographical information are requested by Telegram Bot.

User sends his location information to NodeJs server through Telegram application. The latitude and longitude values are interpreted by the server and create the object for data access. The object and distance metrics are passed to query Firebase database to fetch all the SGBs available in nearby working zone. The circular working zone is characterized with constant radius of 3 km with center point as the latitude and longitude value sent from Telegram application. The radian value is obtained by dividing distance metric with radius value.

Similar to geographical information, transaction details is also stored as separate collection in Firebase database, where it holds details like user name, account number, amount of waste disposed, SGB details, rate for waste, transfer of perCoin rom user account to the SGB account.

The Firebase real-time database is not required for an application server. Security and data endorsement are open through the Firebase real-time database security rules, and rules are composed and executed at the time of data retrieval.

Two information models are used to represent SGB and user transaction details in Firebase database. SGB model stores information related to location and waste. Similarly, the transactions model stores the information about exchanges made in the framework utilizing various traits like client account, weight of waste, cost of transaction, and so on.

Database schemas used in waste management system is detailed in the following sections.

12.4.1.10   SGB Model

Few assumptions are made to simulate SGB in TAG. When users initiate a request to dispose the waste through Telegram application, a node server accepts the request and returns geographical location and waste amount available in SGB. Moreover, it also returns the type of waste and the maximum capacity of the bin as well. Suppose if a user has no balance in account, server adds up minimum balance to SGB owners account. All the transactions are made secured using cryptocurrencies and finally transferred to SGB owner's account.

12.4.1.11   Transaction Model

The transaction model is composed of two different entities for its transaction process:

1. SGB
2. User

User entity holds the information about the disposed waste and payment services, whereas the SGB attribute in the request is bound to particular transaction and each user record is identified using user account number. Similarly, QR code is unique for each transaction, and it helps the bank to distinguish different payment services. When a user wants to dispose of the waste, a server generates QR code, it sends the code to the user. Once he/she receives the QR code, it is scanned and decoded. Finally, the decoded value is sent to server. The server queries the db with the decoded value to check its existence. Based on the waste quantity and type, the amount is transferred from a user account to the SGB owner account, and here all the transactions are carried out perCoin basis.

12.4.1.12 User Domain

Users can interact with TAG in two different modes and they are as follows:

* Bots-Telegram application
* Web application

Users can use Telegram bots and send HTTP requests/commands to interact with TAG application. In addition to bots, users can also use web applications where it affords different approaches to handle and minimize the waste. It also let the users to view the summary of waste disposal, transaction details and so on. Moreover, it also encourages the user to reduce the waste in a liable way.

12.4.1.13 Tools and Technology

Different tools and technologies are used to build overall architecture of smart waste management system. MQTT queuing techniques are used to separate different category messages. Sub categories are introduced between SGB and server to identify the type of the message (i.e., to uniquely identify request/response). Geth and Truffle are utilized to run Ethereum hub and communicate smart contracts to Ethereum blockchain independently.

The first level MQTT hierarchy is classified into SGB and server, whereas the next sub level classification is performed based on the actions required. Thus, MQTT publish/subscribe messages are used for authentication and waste disposal. EthereumNodes are configured and a Geth command line tool is used to run such nodes on any machine environment. Truffle is used as development framework to build and run Ethereum. In this environment, contracts can be compiled, tested, and deployed. The implementation and analysis is carried out for both local and remote nodes with appropriate configurations.

12.4.1.14 Process Flow

The entire process flow of waste management implementation is shown in Figure 12.6.Waste collection is a key functional aspect in the waste management process. The proficient method of collecting waste prompts cleanliness and better wellbeing. Depending on the population and socioeconomic and environmental factors, each country adopted their own rules and regulations to collect the waste in both domestic and industrial sectors. In many cities in the developed countries, waste collectors gather the waste at a fixed time on scheduled day. However, in Korea the cleanliness is highly prioritized, except weekend collector's visit every other day of the week to collect the waste. Waste collectors visit is considered as unserviceable when there isn't sufficient waste to be collected. Subsequently, either cleanliness or eco-friendliness and assets are undermined. The important aspect is

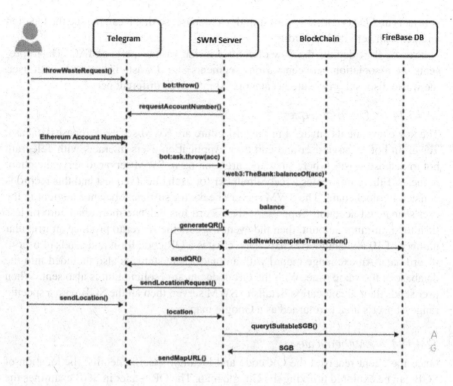

FIGURE 12.6 Process flow.

to identify the optimal route based on the waste status and update it to the concerned division and hence better planning can be made and services can be provided on-time. Since our proposed system is related with the cloud, the statuses of every single waste container all through the city or even nation are available from the cloud. Every one of the stakeholders along with recycling organizations can chart out the plan accordingly.

A most fundamental and simple combination of IoT gadgets with AI and block-chain is presented here.

Despite the fact that blockchain have enormous potential and hence its innova-tions in different dimensions need to be acknowledged. Because of the high process-ing time, blockchain innovation is not efficient for the applications that require fast response time. In order to overcome the dark side of blockchain, TAG has planned hybrid framework, where a central server, named smart waste management server is used to handle the requests that need quick read compose and for the rest, blockchain innovation is used to deal with the financial exchanges.

In this sort of arrangement, the sensors send information to a RaspberryPi and program compiled in RaspberryPi. Further, it uses RPC interface to communicate with the network. Each SGB runs a full Ethereum hub or a private hub is utilized to keep all these SGBs to run in a solitary private system. Every one of the information like client, installment, exchange, SGB is put away in the blockchain. The smart contracts and DAO addresses are stored in database, so that it can pursue the location and perform activities accordingly.

Process flow depicts the flow of control inside various parts of TAG. This represents the association between various segments and if waste is available with user, the waste disposal procedure is categorized into three different parts.

12.4.1.14.1 DisposalRequest

The key components engaged in this procedure are SWM server, Firebase database, Telegram bot to serve user and end user. Application users interacts with Telegram bot to dispose waste, where such bots are running in SWM server to serve the client requests. Ethereum creates a record number for each client request and this record is enlisted in blockchain. The SWM server checks for sufficient balance amount in the user's financial account. Suppose, if the record has balance more than zero or less than the minimum amount, then the event triggers the server to produce an irregular number of 10 alphanumeric characters, creates a QR code for it and sends to user to record data. An exchange record with an inadequate status is also included into the database at the same time. With the QR code, an area solicitation is also sent. When user sends their area location details to SWM server, then all the SGB with a specific range of user's area is returned as a Google map.

12.4.1.14.2 Authenticate

Once the client received the QR code and location specific details, the location of SGB can be explored utilizing the Google map. The QR reader in SGB examines the QR code of user received by Telegram application. When the QR code is checked and decoded, the SGB distributes the decoded code to a MQTT broker to subscribe the topic. The SWM server also checks for the legitimacy of the code and respond to it. Once SWM server recognizes that, it can check the decoded value and bin lids are opened for further process.

12.4.1.14.3 DisposeWaste

Once the cover of SGB is opened, the client can dispose the waste into the SGB based on the communication with the SWM in the above procedures. When the top of SGB is shut, the complete load of the SGB is determined and distributed to the SWM server. The SWM server evaluates the distinction between the past weight and the present load in SGB and finds out the measure of weight client kept. Utilizing weight-rate-based estimation, the amount that needs to pay to the user is determined. The exchange record with inadequate data like expense of administration, client account, date, measure of waste are set as incomplete records in database. Finally, amount needs to be paid to the transaction request initiated by SGB owner and it is made to blockchain using the Web3 interface.

12.4.1.14.4 Route Optimization

When the collection of waste is being done, the waste collectors need to design a superior and eco-friendly route plan by considering the status of waste containers in a city region. By this way, redundant visits are avoided and hence, resources are conserved. Similarly, based on the bin status, trucks are assigned to collect wastes from the local workplaces. The information will be sent to Central Control Center (CCC), where payment calculation is made and priority table is constructed based on waste amount.

The regional sites that are filled up waste are treated first. The highest amount of waste collected by a truck is considered, where if the amount of waste is more than truck carry limit, the algorithm will assign another truck to balance the workload. Multilayer perceptrons is one of the most profound learning methods for nonlinear classification and regression model, and it is often used for classification and prediction. These techniques were used to predict the waste in certain areas and schedule a route plan accordingly.

12.5 CHALLENGES

Though IoT, AI, and blockchain technologies are gaining popularity, development of a real-time system is underway. One of the major downsides is that building a framework whose whole database is running over blockchain consumes more resources and requires more processing time. Moreover, it consumes more electricity to mine the data in the blockchain and hence it always increases complexity for real time implementation. The compilation of IoT data is not compatible where they are different in time and space. Moreover, the incessant flow of IoT data makes the system complicated and increases its complexity. The efficacy of AI depends on the huge volume of data, where it helps them to learn and to make intelligent decisions. Though IoT data serves as feast for AI to learn and enhance its prediction, it pulls down the overall network performance. Lack of knowledge in these emerging technologies among common population is another drawback, where it requires extensive training.

12.6 CONCLUSION AND FUTURE PROSPECTS

In this chapter, a smart waste management system based on three different emerging technologies IoT, AI, and blockchain was proposed. The structural design of a smart waste bin is recreated to help and comprehend the necessity for sending this sort of waste management framework for smart cities. The blockchain innovation was demonstrated to deal with micropayments with least exchange overhead. The micropayments are streamlined based on the amount of waste collected from the bins of each user. The proper disposal of waste helps in sorting and recycling process. To enhance such process, in this smart waste management system user could be remunerated by recycling industries. It depends on the quality and kind of waste they produce. Further, it motivates them to sort and check the waste before they dispose. A reward system was implemented using blockchain smart contracts where it can handle the duplicate payments.

Further, security of the system can be enhanced to take necessary steps to punish the intruders of the system to avoid the exploitation. Similarly, penalty system can also be introduced in future when the quality of waste is below threshold. In penalty system, application notifies the user about the penalty and it expels them to use the system for waste disposal. A large number of applications are built on top of blockchain is keep on increasing and with increasing network efficiency, the demand for blockchain may grow in the future.

REFERENCES

1. Hoornweg, D., and Bhada-Tata, P. (2012). What a waste: A global review of solid waste management. Urban development series; knowledge papers no. 15. World Bank, Washington, DC. © World Bank. https://openknowledge.worldbank.org/handle/10986/17388. License: CC BY 3.0 IGO.

2. Joshi, R., Ahmed, S., and Ng, C. A. (2016). Status and challenges of municipal solid waste management in India: A review. *Cogent Environmental Science*, 2(1), 1–18.

3. Jowit, J. (2010). Recycling still the most effective waste disposal method. *The Guardian*. https://www.theguardian.com/environment/2010/mar/16/recycling-waste-disposal.

4. Komninos, N. (2002). *Intelligent Cities: Innovation, Knowledge Systems, and Digital Spaces*. Spon Press.

5. Wyld, D. C. (2010). Taking out the trash (and the recyclables): RFID and the handling of municipal solid waste. *International Journal of Software Engineering & Applications* (IJSEA), 1(1), 1–13.

6. Jha, S. K. (2014). Smart cities—Solution for Indian Cities. doi:10.13140/2.1.3483.1044. Smart Cities for India, At Delhi NCR.

7. Ganguly. (2016). E-Waste management in India: An overview. *International Journal of Earth Sciences and Engineering*, 9(2), 574–588.

8. Sharholy, M., Ahmad, K., Mahmood, G., Trivedi, R. C. (2008). Municipal solid waste management in Indian cities–A review. *Waste Management*, 28(2), 459–467.

9. Mitton, N., Simplot-Ryl, D., Voge, M.-E., and Zhang, L. (2012). Energy efficient k-anycast routing in multi-sink wireless networks with guaranteed delivery, I, ser. ADHOC-NOW'12, Belgrade, Serbia, pp. 385–398.

10. Prassler, E., Stroulia, E., and Strobe, M. (1997). Office waste cleanup: An application for service robots. *International Conference on Robotics and Automation, (April)*, pages 1863–1868. IEEE.

11. Awasthi, A. K., Wang, M., Wang, Z., Awasthi, M. K., and Li, J. (2018). E-waste management in India: A mini-review. *Waste Management & Research*, 36(5), 408–414.

12. Dwi Atmanti, H., Dwi Handoyo, R., and Muryani. (2018). Strategy for sustainable solid waste management in central Java Province, Indonesia. *International Journal of Advances in Scientific Research and Engineering*, 4(8), 215–223.

13. Lazaro, E. J. P., Alexis, E., and Rubio, J. M. (2014). Solar powered electronic trash can. 2(5), 33–37.

14. Bashir, A., and Banday, S. A. (2013). Concept, design and implementation of automatic waste management system. *International Journal on Recent and Innovation Trends in Computing and Communication*, 604–609.

15. Dugdhe, S., Shelar, P., Jire, S., and Apte, A. (2016). Efficient waste collection system. 2016 International Conference on Internet of Things and Applications (IOTA), 143–147.

16. Thakker, S. (2015). Smart and wireless waste management. *2015 International Conference on Innovations in Information, Embedded and Communication Systems (ICIIECS)*.

17. Guerrero, L. A., Maas, G., and Hogland, W. (2013). Solid waste management challenges for cities in developing countries. *Journal of Waste Management*, 33(1), 220–232.

18. Caniato, M., Vaccari, M., Visvanathan, C., and Zurbrügg, C. (2014). Using social network and stakeholder analysis to help evaluate infectious waste management: A step towards a holistic assessment. *Waste Management*, 34(5), 938–951.

19. Yang, R. J. (2014). An investigation of stakeholder analysis in urban development projects: Empirical or rationalistic perspectives. *International Journal of Project Management*, 32(5), 838–849.

20. Greene, K. L., and Tonjes, D. J. (2014). Quantitative assessments of municipal waste management systems: Using different indicators to compare and rank programs in New York State. *Waste Management*, 34(4), 825–836.

21. Kollikkathara, N., Feng, H., and Stern, E. (2009). A purview of waste management evolution: Special emphasis on USA. *Waste Management*, 29(2), 974–985.

22. Zhang, X., and Huang, G. (2014). Municipal solid waste management planning considering greenhouse gas emission trading under fuzzy environment. *Journal of Environmental Management*, 135, 11–18.

23. Moh, Y. C., and Manaf, L. A. (2014). Overview of household solid waste recycling policy status and challenges in Malaysia. *Resources, Conservation and Recycling*, 82, 50–61.

24. Nagasubramanian, G., Sakthivel, R. K., Patan, R., Gandomi, A. H., Sankayya, M., and Balusamy, B. (2018). Securing e-health records using keyless signature infrastructure blockchain technology in the cloud. *Neural Computing and Applications*, 32, 639–647.

25. Kumar, S. R., Gayathri, N., Muthuramalingam, S., Balamurugan, B., Ramesh, C., and Nallakaruppan, M. K. (2019). Medical big data mining and processing in e-healthcare. In *Internet of Things in Biomedical Engineering*, pp. 323–339. Academic Press, London, UK.

26. Dhingra, P., Gayathri, N., Kumar, S. R., Singanamalla, V., Ramesh, C., and Balamurugan, B. (2020). Internet of Things–based pharmaceutics data analysis. In *Emergence of Pharmaceutical Industry Growth with Industrial IoT Approach*, pp. 85–131. Academic Press, London, UK.

27. Al-Jarallah, R., and Aleisa, E. (2014). A baseline study characterizing the municipal solid waste in the State of Kuwait. *Waste Management*, 34(5), 952–960.

28. Bing, X., Bloemhof, J. M., Ramos, T. R. P., Barbosa-Povoa, A. P., Wong, C. Y., and van der Vorst, J. G. (2016). Research challenges in municipal solid waste logistics management. *Waste Management*, 48, 584–592.

29. Yang, G. T. M., and Thung, G. (2016). Classification of trash for recyclability status. CS229 Project Report, vol. 2016.

30. Arebey, M., Hannan, M., Begum, R. A., and Basri, H. (2012). Solid waste bin level detection using gray level cooccurrence matrix feature extraction approach. *Journal of Environmental Management*, 104, 9–18.

31. Simonyan, K., and Zisserman, A. (2014). Very deep convolutional networks for large-scale image recognition. *arXiv preprint* arXiv:1409.1556.

32. Sinha, A., and Couderc, P. (2012). Using owl ontologies for selective waste sorting and recycling. In *OWLED-2012*.

33. (a) Thomas, V. M. (2008). RFID helps reward consumers for recycling. *RFID Journal*. (b) Thomas, V. M. (2008). Environmental implications of RFID. In *Electronics and the Environment, ISEE 2008. IEEE International Symposium on*, pages 1–5. IEEE.

34. Thung, G., and Yang, M. (2016). Classification of trash for recyclability status. *arXiv Preprint*.

35. Rad, M. S., Kaenel, A. V., Droux, A. et al. (2017). A computer vision system to localize and classify wastes on the streets. In *Computer Vision System*, pp. 195–204. Springer, Cham, Switzerland.

36. Awe, O., Mengistu, R., and Sreedhar, V. (2017). Smarttrashnet: Waste localization and classification. *arXiv Preprint*.

37. He, K., Zhang, X., Ren, S., and Sun, J. (2016). Deep residual learning for image recognition. In *Proceedings of Conference on Computer Vision and Pattern Recognition*. Las Vegas Valley, NV.

38. Krizhevsky, A., Sutskever, I., and Hinton, G. E. (2012). ImageNet classification with deep convolutional neural networks. In *Proceedings of Neural Information Processing System Conference*. Lake Tahoe, CA.

39. Sermanet, P., Eigen, D., Zhang, X., Mathieu, M., Fergus, R., and Lecun, Y. (2014). OverFeat: Integrated recognition, localization and detection using convolutional networks. In *Proceedings of International Conference on Learning Representations*. Banff, Canada.

40. Zeiler, M. D., and Fergus, R. (2014). Visualizing and understanding convolutional neural networks. In *Proceedings of 13th European Conference on Computer Vision (ECCV)*. Zurich, Switzerland.

41. Donovan, J. (2018). Auto-trash sorts garbage automatically at the techcrunch disrupt hackathon. Available https://techcrunch. com/2016/09/13/auto-trash-sorts-garbage-automatically-atthe-techcrunch-disrupt-hackathon.

42. Ravale, U., Patel, N., and Khade, A. (2017). Smart trash: An efficient way for monitoring solid waste management, *2017 International Conference on Current Trends in Computer, Electrical, Electronics and Communication (CTCEEC)*.

43. Pathak, P., and Srivastava, R. R. (2017). Assessment of legislation and practices for the sustainable management of waste electrical and electronic equipment in India. *Renewable and Sustainable Energy Reviews*, 78, 220–232.

44. Abdoli, S. (2009). RFID application in municipal solid waste management system. *International Journal of Environmental Research*, 3(3), 447–454.

45. Altman, N. S. (1992). An introduction to kernel and nearest-neighbor nonparametric regression. *The American Statistician*, 46(3), 175–185.

46. Antonisamy, B., Premkumar, P. S., and Christopher, S. (2017). *Principles and Practice of Biostatistics-E-book*. Elsevier Health Sciences, New Delhi, India.

47. Arebey, M., Hannan, M., Basri, H., Begum, R. A., and Abdullah, H. (2011). Integrated technologies for solid waste bin monitoring system. *Environmental Monitoring and Assessment*, 177(1–4), 399–408.

48. Álvarez-Chávez, C. R., Edwards, S., Moure-Eraso, R., and Geiser, K. (2012). Sustainability of bio-based plastics: General comparative analysis and recommendations for improvement. *Journal of Cleaner Production*, 23(1), 47–56.

49. Chowdhury, B., and Chowdhury, M. U. (2007). Rfid based real-time smart waste management system. In *Telecommunication Networks and Applications Conference*, pages 175–180. IEEE.

50. Cortes, C., and Vapnik, V. (1995). Support-vector networks. *Machine Learning*, 20(3), 273–297.

51. Glouche, Y., and Couderc, P. (2013). A smart waste management with self-describing objects. In *The Second International Conference on Smart Systems, Devices and Technologies (SMART'13)*.

52. Ho, T. K. (1995). Random decision forests. In *Document Analysis and Recognition, Proceedings of the 3rd International Conference on*, volume 1, pages 278–282. IEEE.

53. Islam, M. S., Arebey, M., Hannan, M., and Basri, H. (2012). Overview for solid waste bin monitoring and collection system. In *Innovation Management and Technology Research (ICIMTR), 2012 International Conference on*, pages 258–262. IEEE.

54. Krizhevsky, A., Sutskever, I., and Hinton, G. E. (2012). Imagenet classification with deep convolutional neural networks. In *Advances in Neural Information Processing Systems*, pages 1097–1105.

55. Geirhos, R., Rubisch, P., Michaelis, C., Bethge, M., Wichmann, F. A., and Brendel, W. (2018). Imagenet-trained cnns are biased towards texture; increasing shape bias improves accuracy and robustness, *arXiv preprint* arXiv:1811.12231.

56. Costa, B. S., Bernardes, A. C. S., Pereira, J. V. A., Zampa, V. H., Pereira, V. A., Matos, G. F., Soares, E. A., Soares, C. L., and Silva, A. (2018). Artificial intelligence in automated sorting in trash recycling. *Anais do XV Encontro Nacional de Inteligência Artificial e Computacional*, pp. 198–205. doi:10.5753/eniac.2018.4416.
57. Ren, S., He, K., Girshick, R., and Sun, J. (2017). Faster RCNN: Towards real-time object detection with region proposal networks. *IEEE Transactions on Pattern Analysis and Machine Intelligence*, 39(6), 1137–1149. doi:10.1109/TPAMI.2016.2577031.
58. Chaudhary, K., Mathiyazhagan, K., and Vrat, P. (2017). Analysis of barriers hindering the implementation of reverse supply chain of electronic waste in India. *International Journal of Advanced Operations Management*, 9(3), 143–168.
59. Shahabdeen, J. A. Smart garbage bin, June 23, 2016. US Patent App. 14/578,184.
60. Russakovsky, O., Deng, J., Su, H., Krause, J., Satheesh, S., Ma, S., Huang, Z. et al. (2015). Imagenet large scale visual recognition challenge. *International Journal of Computer Vision*, 115(3), 211–252.
61. Parlikad, A. K., and McFarlane, D. (2007). RFID-based product information in end-of-life decision making. *Control Engineering Practice*, 15(11), 1348–1363.
62. Kumar, S. R., and Gayathri, N. (2016). Trust-based data transmission mechanism in MANET using sOLSR. In *Annual Convention of the Computer Society of India*, pp. 169–180. Springer, Singapore.
63. Muthuramalingam, S., Bharathi, A., Gayathri, N., Sathiyaraj, R., and Balamurugan, B. (2019). IoT Based Intelligent Transportation System (IoT-ITS) for Global Perspective: A Case Study. In *Internet of Things and Big Data Analytics for Smart Generation*, pp. 279–300. Springer, Cham, Switzerland.

13 Quantum Computation for Big Information Processing

Tawseef Ayoub Shaikh and Rashid Ali

CONTENTS

13.1 INTRODUCTION

Big data (BD), a topic of hot discussion embodying almost all circles, arrived with a fervor that parallels some of the most significant movements in the history of computing. Data nowadays is produced at an exponentially growing rate, which puts a big challenge for the analysis of this mighty resource and limits possibilities for knowledge to be gained. Digital media devices have already been more than

92% in 2002 [1], while the size of this new data was also more than 5 exabytes (5×10^{18} bytes). It puts a huge contest to analyze such gigantic scale data even though computer systems today are much faster than those in the 1930s. Every day 2.5 exabytes of data is produced in the world with the bang in digital technologies, making it equivalent to the memory of 5 million laptops or 150 million phones. Quite a few efficient methods such as sampling, data condensation, density-based approaches, grid-based approaches, divide and conquer, incremental learning, and distributed computing, have been presented in response to the problems of analyzing large-scale data [2]. Of course, these methods are constantly used to improve the performance of the operators in analytics process, the results of these methods illustrate that with the efficient methods at hand, we may be able to analyze the large-scale data in a reasonable time. Principal component analysis (PCA) [3], a dimensional reduction method, is a typical example that is aimed to accelerate the process of data analytics by reducing the input data volume. Sampling [4], which can also be used to speed up the computation time of data analytics, is another reduction method that reduces the data computations of data clustering. Following Moore's law for several decades, although the advances of computer systems and internet technologies have witnessed the development of computing hardware, the problems of handling large-scale data still exist when we are entering the age of big data. An intelligent way of processing, organizing, and extracting hidden value from gigantic data piles like big data is desired since complexity, size of data sets upswings balloons year after year. Quantum computing is the answer. Big data and machine learning have strapped the limits of current information technology (IT) infrastructure for handling large troves of datasets effectually which gave birth to the fresh and exciting archetype of quantum computing (QC) that has the power to dramatically upsurge the speed [5]. Quantum computing permits for quick detection, analysis, integration, extensive search and diagnosis from bulky sprinkled data sets to quickly uncover patterns. Quantum computers will be able to complete complex calculations in mere seconds—same calculations that would take today's computers thousands of years to solve [6].

Encouraging developments in the quantum computing direction have already started. Open quantum assembly language (QASM) and software development kits (SDKs) like quantum information science kit (QISKit) are fresh languages that open new horizons for the research and development in the new paradigm of quantum computing. Open Fermion, an open-source quantum computing platform, was launched by Google in October 2017. It may bring revolutions in the dimensionality reduction of input and output data of the Harrow-Hassidim-Lloyd (HHL: a quantum algorithm for linear systems of equations) algorithm [7]. The restricted computing power of prevailing computers has encountered bottlenecks in the application of quantum computing in quantum simulation machine learning and the number and quality of data processed are faced with many problems. The quantum computing and quantum algorithms emergence may become an effective means for disruption of these bottlenecks. Also, quantum principal component analysis (PCA) is more efficient than traditional PCA exponential. A good wealth of quantum computing algorithms could be found from the literature like Grover's algorithm [7], The Deutsch–Jozsa algorithm [8–10], Shor's algorithm [11], Simon's algorithm [12], Quantum Fourier transform (QFT) [13]. Ying et al. [14] discuss some potential meeting points between quantum computing and artificial intelligence (AI) as:

- Quantum algorithms for learning (QAL)
- Quantum algorithms for decision-making (QAD)
- Quantum search (QS)
- Quantum game theory (QGT)
- Semantic analysis (SA)
- Natural language (NL)
- Quantum Bayesian networks (QBN)
- Quantum neural networks (QNN)
- Quantum genetic algorithms (QGA)

13.2   WHAT DOES THAT MEAN FOR BIG DATA?

The progress in these fields critically relies on processing power. The computational requirement of big data analytics (BDA) is currently placing a considerable strain on computer systems. Since 2005, the focus has been shifted to parallelism using multiple cores instead of a single fast processor. However, many problems in big data cannot be solved simply by using more and more cores. Splitting up the work among multiple processors is used but its implementation is complex. The problems need to be solved sequentially where the preceding step is equally important. At the Large Hadron Collider (LHC) in the European Council for Nuclear Research (CERN), Geneva, particles are accelerated, traveling at almost the speed of light within a 27 km ring such that 600 million collisions take place in a second wherein only one of the 1 million collisions is chosen for pre-selection. In the pre-selection process, only 1 out of 10,000 events are passed to a grid of processor cores that further choose 1 out of 100 possible events, hence, making the data process at 10 GB/s. At LHC, 5 trillion bits of data are captured every second and after discarding 99% of the data, it still analyzes 25 petabytes of data a year.

Such is the power of quantum computing but the current resources lack the potential for its possible application in big data. If it were possible, the computing would be useful for specific tasks such as factoring large numbers that are useful in cryptography, weather forecasting, searching through large unstructured data sets in a fraction of the time, identification of patterns and anomalies, etc. The developments in quantum computing could actually make encryption obsolete in a jiffy. With such computing powers, it would be one day possible to make large datasets that would probably store complete information such as the genetics of every single human that existed, and machine learning algorithms could find patterns in the characteristics of these humans while also protecting the identities of the humans. Also, clustering and classification of data would become a much faster task. The future says quantum computers will allow faster analysis and integration of our enormous data sets, which will improve and transform our machine learning and artificial intelligence capabilities.

Quantum computing seems to have exceptional promise in big data analytics in the future. Fruitful work in the recent few years has been carried out on applications of quantum computing on big data analytics. The present work may be considered as an initiative in the same direction. The main contribution of the paper is to present the study in the area of quantum computing for big data analytics focusing mainly on the healthcare sector. Also, we present a model explaining the way out of how quantum computing can tackle the complex issues presented by big data which will become a platform for further research in the same challenging field.

13.3   DEFINITION OF BIG DATA

Big data does not have a single definition. International Business Machines (IBM) defines it as the data characterized by three Vs (3Vs), namely volume, variety, and velocity. There has been the addition of a few more Vs to this definition like veracity, value, and variability.

1. Volume (size): It is the scale and size of the data available today. By the modern hi-fi technologies like instrumentation, we are now able to sense more things by which people and things are getting increasingly interconnected. This interconnectivity rate is a runaway train, generally referred to as machine-to-machine (M2M) interconnectivity and is responsible for double-digit year over year (YoY) data growth rates. These mountains of data pulps yield a lot of opportunity for useful insights but parallel comes with an open challenge as it is said, "challenge is the biggest opportunity."

2. Variety (complexity): It refers to the complexity of the data. Initially, data was stored in tables like relational tables, which were having a predefined schema. But with the data available from diverse sources and data in motion (streams), it is the utmost need to integrate these diverse data formats so as derive the productive knowledge, which is not possible from a single source of data. It is humbling to see that we spent more of our time on just 20% of the data and pay less attention to 80% of the world's data, leading into a blind zone as depicted in Figure 13.1. To capitalize on the big data opportunity, enterprises must be able to analyze all types of data, both relational and non-relational: text, sensor data, audio, video, transactional, and more.

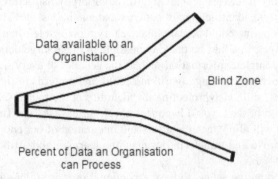

FIGURE 13.1   Percentage of data available to an organization vs. data they actually use.

3. Velocity: It is the speed of creating, storing, and analyzing the data. Since the big data applications need the real-time processing of the data for action-based results so the rate needed for data-driven actions should be compatible with the rate of generation and processing of the data. Dealing effectively with big data requires to perform analytics against the volume and variety of data while it is still in motion, not just after it is at rest.

4. Veracity: This is the parameter used to measure the quality, validity, and volatility so as to be sure about the accuracy of data, reliability of the data source, the context within the analysis.

5. Valance: It is the connectedness. The more the data is connected, the more is the valance. Since the valance increases over time, it makes the data connections denser.

6. Value: It is the integration of all the different formats of the data in order to get valuable insights from it in the form of value.

The rest of this chapter is organized as follows: Section 13.4 deals with the exploration of related work available from the literature sources followed by a respective review of each machine learning algorithm from recent existing literature. Firstly, the categorization of different machine learning techniques with their quantum versions is prepared followed by a brief review of each one of them. Section 13.5 focuses on a detailed description of a few important types of quantum gates with their respective use in solving various types of computation problems. Section 13.6 discusses broadly the awesome properties like entanglement, interference and quantum parallelism. Section 13.7 throws detailed light on the rising mountains' amount of complex data and parallelly provides a glimpse of how those issues can be resolved by the quantum machine learning algorithms. The proposed methodology and detailed discussion are carried in Section 13.8, in which a model through a diagram is presented to highlight how quantum machine learning algorithms can make the healthcare smart. It also describes the potential of quantum computing in big data analytics and also respective challenges to harness its power in big data analytics. Finally, conclusion about the whole work is done through summarization in addition to the depiction of some future, in Section 13.9. Finally, it ends with the references section.

13.4   RELATED WORK

The theme of this section is the exploration of related work available from the literature sources. We have subdivided this work into five categories followed by an extensive survey in each group. In this section, we have categorized the different machine learning techniques with their quantum versions followed by a brief review of each one of them. Firstly, it is quantum classification which is subdivided again into quantum K-means, support vector machines, and regression. Similarly, clustering consists of three tastes of quantum versions namely quantum K-means, K-medians, and quantum hierarchical clustering.

Quantum algorithms for machine learning are using the properties of elementary particles to find patterns in data. Grover's search [7], an algorithm to find elements in an unordered set quadratically faster than by any classical variant, provides the basic cornerstone upon which many quantum learning algorithms rely on. Mostly unsupervised learning [15] methods use this approach: *K*-medians [16], hierarchical clustering, and quantum manifold embedding. In addition, quantum neural networks and quantum associative memory often rely on this search and also an early version of quantum support vector machines. Machine learning is a branch of artificial intelligence, gains its power by learning from previous experiences in order to predict the future by making reasonable decisions, offering exceptional opportunities in fields like computer sciences, bioinformatics, financial analysis, and robotics. The present world of big data creates a challenge to the machine learning with the pace of incrementally growing rate of "big data" that could become intractable for classical computers. The exponential speedup is possible in scenarios where both the input and the output are also quantum. Listing class membership or reading the classical data, one would imply at least linear time complexity, which could only be a polynomial speedup. Examples include quantum principal component analysis, quantum *K*-means and a different flavor of quantum support vector machines. Regression, based on quantum process tomography requires an optimal input state and in this regard, it needs a quantum input [17]. At a high level, it is possible to define an abstract class of problems that can only be learned in polynomial time by quantum algorithms using quantum input [18].

13.4.1 QUANTUM MACHINE LEARNING

For big data classification, Rebentrost et al. [19] offered a quantum-based support vector machine with a time complexity O (log NM), where N is the number of dimensions and M is the number of training data. Big data mining has bright visions by using a quantum-based search algorithm when maturity takes the quantum computing field. Drawing on Trugenberger's proposal for measuring the Hamming distance on a quantum computer and discuss its advantages using handwritten digit recognition from the Modified National Institute of Standards and Technology (MNIST) database. M. Schuld et al. [20] introduce a quantum pattern classification algorithm that takes a computationally expensive subroutine from an inventive machine learning algorithm and executing it more professionally on a quantum computer. They concluded from the results that a quantum computer takes time O (log (MN)) as compared with time O (poly (MN)) for the best known classical problem of assigning N-dimensional vectors [21–24]. For the classification of mysterious quantum states, authors [25,26] et al. endeavored the intuitions of Bayesian decision theory. Solving learning optimization problem using adiabatic quantum computing is put forward by authors [27–29] et al. Also, the nearness of open quantum systems to machine learning methods grounded on Markov models is underlined by authors [30,31] et al.

Quantum computation machine learning is evolving as an authoritative tactic letting improvements in speed-ups and classical machine learning algorithms. The drop-in prerequisite time for training a restricted Boltzmann machine (RBM), while also providing a wealthier framework for deep learning than its classical colleague is offered by authors et al. [33,34].

Shaikh et al. [35] reviewed the available literature on big data analytics (BDA) using quantum computing (QC) for machine learning (ML) and its current state of the art by categorizing quantum machine learning in different subfields depending upon the logic of their learning followed by a review in each technique. Quantum walks is used to construct quantum artificial neural networks, which exponentially speed up the quantum machine learning algorithm is discussed. Quantum supervised and unsupervised machine learning and its benefits are compared with that of the classical counterpart.

13.4.1.1 Classification

13.4.1.1.1 K-Nearest Neighbors

In the K-nearest neighbor's algorithm [36], calculations only take place when a new, unlabeled instance is presented to the learner. On seeing the new instance, the learner searches for the K most nearby data instances [37]. This is the computationally expensive step that is easily accelerated by quantum methods. Among those K instances, the algorithm will select the class which is the most frequent.

13.4.1.1.2 Support Vector Machine

A support vector machine is a supervised learning algorithm that learns a given independent and identically distributed a set of training instances $\{(x1, y1), (and, Z)\}$, where $y \in \{-1, 1\}$ are binary classes to which data points belong [38]. A hyperplane in Rd has the generic form

$$W * x - b = 0 \qquad (13.1)$$

where w is the normal vector to the hyperplane, and the bias parameter b helps determine the offset of the hyperplane from the origin.

The calculations are dominated by the kernel evaluation. The calculation of entry in the kernel matrix takes O(d) time on a linear or polynomial kernel, thus, calculating the whole kernel matrix has O(N^{2}d) time complexity and accordingly solving the quadratic dual problem or the least-squares formulation has O(N^3) complexity. So the classical support vector machine algorithm has at least O (N^2 (N + d)) complexity upon combining the two steps [38]. This complexity can be mitigated by using spatial support structures for the data. The quantum formulation yields an exponential speedup in these two steps, leading to an overall complexity of O (log (Nd)).

13.4.1.1.3 Regression Analysis

In regression, we seek to approximate a function x given a finite sample of training instances $\{(x1, y1),\dots, (xN, yN)\}$, resembling supervised classification [39]. Unlike in classification, however, the range of yi is not discrete as it can take any value in R. The approximating function f is also called the regression function. A natural way of evaluating the performance of an approximating function f is the residual sum of squares:

$$E = \sum_{i}^{N}\left(yi - f\left(xi^2\right)\right) \qquad (13.2)$$

Solving the normal equation in linear least squares regression is typically done via singular value decomposition whose overall cost is O (dN2) steps. Matrix inversion offers an efficient quantum variant, especially if both the input and output are also quantum states. In nonparametric regression, we usually face polynomial complexity. Using the K-nearest neighbor's algorithm, we find the complexity is about cubic in the number of data points. The most expensive step is finding the nearest neighbors, which is done efficiently by quantum methods.

An algorithm that solves the problem of pattern classification on a quantum computer, performing linear regression effectively with least-squares optimization is given by Schuld et al. [40]. It runs in time logarithmic in the dimension N of the feature vectors as well as independent of the size of the training set if the inputs are given as quantum information. Instead of requiring the matrix containing the training inputs X to be sparse, it merely needs X*X to be represented by a low-rank approximation.

Rebentrost et al. [41] used a support vector machine for implementing an optimized linear and non-linear binary classifier on a quantum computer whose speed was exponential in the size of the vectors and the number of training examples. For efficiently performing a principal component analysis and matrix inversion for training the data kernel matrix, lies at the basic core of the algorithm.

In [42] et al., authors discussed the application of quantum computing to solve the problem of effective SVM training especially in the case of digital implementations. A comparison of the behavioral aspects of conventional and enhanced SVMs is carried out and experiments in both synthetic and real-world problems are also carried to support the theoretical analysis. The presented research at the same time differences between quadratic programming and quantum-based optimization techniques.

13.4.1.2 Clustering

13.4.1.2.1 Quantum K-Means

Most quantum clustering algorithms are based on Grover's search. These clustering algorithms offer a speedup compared with their classical counterparts, but they do not improve the quality of the resulting clustering process. This is based on the belief that if finding the optimal solution for a clustering problem is NP-hard, then quantum computers would also be unable to solve the problem exactly in polynomial time. If we use quantum random access memory (QRAM) [43], an exponential speedup is possible. Centroids are being calculated and vectors are assigned to the closest centroids. The simplest quantum version of K-means clustering calculates, like the classical variant, but using Grover's search to find the closest ones. The complexity is O (N log (Nd)) since every vector is tested in each step, which is an exponential speedup over the polynomial complexity of classical algorithms. Further improvement is possible if we allow the output to be quantum. Every algorithm that returns the cluster assignment for each N output must have at least O (N) complexity.

13.4.1.2.2 Quantum K-Medians

In K-means clustering, the centroid may lie $o\left(\frac{N^{3/2}}{\sqrt{K}}\right)$ outside the $o\left(\frac{N^2}{K}\right)$ manifold in which the points are located [44]. K-medians bypass this problem by always choosing an element in a cluster to be the center is a flavor of this family of algorithms. This task is achieved by constructing a routine for finding the median in a cluster using Grover's search and then iteratively calling this routine to reassign elements. The median is identified among a set of points in O ($N\sqrt{N}$) time. Quantum K-medians have complexity as opposed to the classical algorithm.

13.4.1.2.3 Quantum Hierarchical Clustering

Quantum hierarchical clustering hinges on ideas similar to those of quantum K-medians clustering. Here we use a quantum algorithm to calculate the maximum distance between two points in a set instead of finding the median. In order to split clusters and reassign the data instances to the most distant pair of instances, we iteratively call this algorithm. This is the divisive form of hierarchical clustering. Quantum divisive clustering O (N log) has much lower complexity compared with the classical limit O (N^2).

For cluster finding and cluster assignment, Lloyd et al. [20] used supervised and unsupervised quantum machine learning algorithms. The work shows that quantum machine learning can provide exponential speedups over classical computers for a good number of learning tasks. A quantum computer takes time O(log(MN)) as compared with time O(poly(MN)) for the best-known classical during the task of assigning N-dimensional vectors to one of several clusters of M states explaining quantum machine learning can provide an exponential speed-up for problems involving large numbers of vectors as well ("big quantum data").

Using the quantum adiabatic algorithm (a quantum version of Lloyd's) to perform K-means clustering of M vectors into K clusters in time O(log k (MN)) can be classified. Privacy is also enhanced by quantum machine learning that allows only O(log (MN)) calls to the quantum database for cluster assignment in comparison to O(MN) to uncover the actual data. The database user can still obtain information about the desired patterns while the database owner is assured that the user has only accessed an exponentially small fraction of the database.

In [45] et al., the authors used a quantum paradigm for speeding up unsupervised learning algorithms and to accelerate learning algorithms by quantizing some of their subroutines. Quantized versions of clustering via minimum divisive clustering, spanning trees and K-medians that are faster with respect to their classical analogs are given. The work also gives a distributed version of K-medians that allows the participants to save on the global communication cost of the protocol compared to the classical version is. Lastly, quantum algorithms for the construction of a neighborhood graph are given.

13.4.1.3 Optimization

R. Shang et al. [46] present an approach to large-scale CARP called quantum-inspired immune clonal algorithm (QICA-CARP) that chains the feature of quantum computation and artificial immune system grounded on the qubit and the quantum superposition. The convergence rate and the quality of the acquired solutions from experimental results show that QICA-CARP beats other algorithms. R. Xia et al. [47] report a hybrid quantum algorithm hiring a restricted Boltzmann machine to gain accurate molecular potential energy surfaces. For optimizing the primary objective function developing a quantum algorithm, the authors achieved a resourceful technique for the calculation of the electronic ground state energy for a small molecule system. P. Diamandis, et al. [48] described quantum computing promises to model molecular interactions at an atomic level so as to understand various diseases. Using quantum computer simulations can be the way for designing and choosing the next generations of drugs and cancer cures.

13.4.1.4 Scheduling

For the purpose of solving the dynamic flexible job-shop scheduling problem, T. Ning et al. [49] establishes the mathematical model to minimize the makespan and stability value, an improved double chains quantum genetic algorithm using quantum techniques is proposed. K. Qazi et al. [50] proposed a complex-valued quantum neural network approach to predicting workloads in data centers. According to the results, multi-layered neural networks with multi-valued neurons (MLMVN) shows higher accuracy compared to both long short-term memory recurrent neural network (LSTM-RNN) and echo state network (ESN), for all the prediction lengths. J. Cruz-Benito et al. [51] describe a project of IBM Q team that implements an intelligent system based on a deep learning approach that learns how people code using the Open QASM language to later offer help and guidance to the coders by recommending different code sequences, logical steps or even small pieces of code.

13.4.1.5 Privacy

Privacy and cryptography are the main branches in which quantum computing has beared its fruits. Quantum cryptography (QKD) is quite a mature field nowadays in providing a more secure environment for IT use. Useful work and future direction in QKD are available in [52–55]. A detailed discussion about opportunities and challenges of quantum computing in artificial intelligence/machine intelligence is put forward by pioneering work in [56,57]. The implementation of soft computing techniques feasible for quantum computing too is discussed here.

13.5 QUANTUM COMPUTING

This section focuses on a detailed description of a few important types of quantum gates with their respective use in solving various types of computation problems.

Quantum computing is computation carried with the laws of quantum mechanics. Quantum information processing (QIP) [58] is the application of quantum mechanics concepts in the field of information processing, which gains its potential power from quantum resources such as superposition, quantum parallelism, entanglement, and quantum interference. Quantum computation being entirely a new way of information processing has let the traditional methods of computing and information processing be referred to as classical information. The yes-no answer to a question basically represented by a most basic piece of information is called a bit. A two-state system representing information using base 2 or binary numbers are used to represent this mathematically. Since a binary number can be either 0 or 1, so a bit can assume one or the other of these values. Physically, a bit is implemented with an electrical circuit that is either at the ground or zero volts (binary 0) or at say +5 volts (binary 1).

13.5.1 What Is a Qubit?

Qubit is analogous to a classical bit representing the basic unit of information processing used in quantum computation, which stays for a quantum bit [59]. Superficially,
a qubit is going to look in some way similar to a bit, but it is totally fundamentally
different, which allows us to do information processing in new and interesting ways.
Like a bit, a qubit can also be in one of two states labeled by the two states as $|0>$
and $|1>$. $|>$ in quantum theory enclosed by an object can be called a state, a vector,
or a ket. So how is a qubit different from an ordinary bit? While a bit in an ordinary
computer can possess either state 0 or state 1, a qubit is somewhat more general.
A qubit can exist in the state $|0>$ or the state $|1>$, but it can also exist in what we call
a superposition state that is a linear combination of the states $|0>$ and $|1>$ which is
made more clear by Figure 13.2. If we label this state $|\psi>$, a superposition state is
written as:

$$|\psi> = \alpha|0> + \beta|1> \tag{13.3}$$

Here α, β are complex numbers. That is, the numbers are of the form $z = x + iy$, where
$i = \sqrt{-1}$.

While a qubit can exist in a superposition of the states $|0>$ and $|1$, whenever we
make a measurement we aren't going to find it like that. In fact, when a qubit is mea-
sured, it is only going to be found to be in the state $|0>$ or the state $|1>$. The modulus
squared of α, β in gives us the probability of finding the qubit in state $|0>$ or $|1>$,
respectively as governed by the laws of quantum mechanics. In other words:

$|\alpha|^2$: Tells us the probability of finding $|\psi>$ in state $|0>$.
$|\beta|^2$: Tells us the probability of finding $|\psi>$ in state $|1>$.

The fact that probabilities must sum to one puts some constraints on what the mul-
tiplicative coefficients in it can be. The values α and β are constrained so as to let

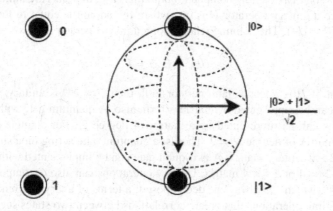

FIGURE 13.2 Classical bit vs. qubit.

the squares of these coefficients get related to the probability of obtaining a given measurement result, by the requirement that:

$$|\alpha|^2 + |\beta|^2 = 1 \qquad (13.4)$$

Generally speaking, if an event has N possible outcomes [59] and we label the probability of finding the result I by pi, the condition that the probabilities sum to one is written as:

$$\sum_{i}^{n} pi = p1 + p2 + p3....pn = 1 \qquad (13.5)$$

The modulus of these numbers can be calculated in the following way:

$$|\alpha|^2 = (\alpha)(\alpha*)$$

$$|\beta|^2 = (\beta)(\beta*)$$

where $\alpha*$ is the complex conjugate of α and $\beta*$ is the complex conjugate of β. We recall that to form the complex conjugate of $z = x + iy$, we let $i \rightarrow -i$. Therefore, the modulus of a complex number z is:

$$|z|^2 = (x + iy)(x - iy) = x^2 + ixy - ixy + y^2 \qquad (13.6)$$

$$\Rightarrow |z| = \sqrt{x^2 + y^2}$$

13.5.2 Single Qubit Gates

A gate is an abstraction that represents information processing. Gates are also used for information in a quantum computer, but in this case, the gates are unitary operations. Since a unitary operator U is one where the adjoint is equal to the inverse, meaning $U^\dagger = U-1$. The defining relation for a unitary operator is thus:

$$UU^\dagger = U^\dagger U = I \qquad (13.7)$$

In addition, if H is a Hermitian operator [60], then $U = e^{iHt}$ is unitary. Quantum operators can be represented by matrices form also. A quantum gate with n inputs and outputs can be represented by a matrix of degree 2^n. For a single qubit, we require a matrix of degree $2^1 = 2$. That is, a quantum gate acting on a single qubit will be a 2×2 unitary matrix. A two-qubit gate can be implemented with a matrix of degree $2^2 = 4$ or a 4×4 matrix. Unitary operations can also decompose two or more qubits, which can always be decomposed in terms of unary and binary gates. The entangling operations that create correlations between two states such that the resulting state cannot be factored into a product of the individual states are also possible. Several algorithms, including integer factorization, unstructured search and the simulation of quantum many-body systems have been shown to be more efficient using qubits.

$$|0>= \begin{pmatrix} 1 \\ 0 \end{pmatrix}, |1>= \begin{pmatrix} 0 \\ 1 \end{pmatrix} \tag{13.8}$$

The Pauli X matrix, which we will often refer to as the NOT operator in case of the qubit, is given in matrix form in the standard or computational basis as:

$$X = UNOT = \begin{pmatrix} 0 & 1 \\ 1 & 0 \end{pmatrix} \tag{13.9}$$

Hence, we have,

$$UNOT \ |0>= \begin{pmatrix} 0 & 1 \\ 1 & 0 \end{pmatrix}\begin{pmatrix} 1 \\ 0 \end{pmatrix} = \begin{pmatrix} 0 \\ 1 \end{pmatrix} = |1> \tag{13.10}$$

$$UNOT \ |1>= \begin{pmatrix} 0 & 1 \\ 1 & 0 \end{pmatrix}\begin{pmatrix} 0 \\ 1 \end{pmatrix} = \begin{pmatrix} 1 \\ 0 \end{pmatrix} = |0> \tag{13.11}$$

So with respect to the standards or computational basis, the X matrix acts as a NOT operator. The action of a NOT gate on an arbitrary state $|j>$ can be written using the XOR operation as:

$$X \ |j>= \ |j> \oplus |1> \tag{13.12}$$

13.5.3 Hadamard Gates

An important gate in quantum computation is the Hadamard gate [61], which is used to create superposition states. Hadamard gate H acts on the computational basis states in the following way:

$$H|0>= \frac{|0>+|1>}{\sqrt{2}}, \ H|1>= \frac{|0>-|1>}{\sqrt{2}}, \tag{13.13}$$

Two Hadamard gates in the series act to reverse the operation and give back the original input resulting in an interesting feature of the Hadamard gate. Let's consider a Hadamard gate applied to an arbitrary qubit $|\psi>=\alpha|0>+\beta|1>$, we have

$$H \ |\psi >= \alpha H \ |0> +\beta H \ |1>= \alpha \left(\frac{|0>+|1>}{\sqrt{2}} \right) +\beta \left(\frac{|0>-|1>}{\sqrt{2}} \right)$$

$$= \left(\frac{\alpha+\beta}{\sqrt{2}} \right)|0> \left(\frac{\alpha-\beta}{\sqrt{2}} \right)|1> \tag{13.14}$$

If we apply a Hadamard gate twice, we get the original state back:

$$H\left[\left(\frac{\alpha+\beta}{\sqrt{2}}\right)\right]|0> + \left[\left(\frac{\alpha-\beta}{\sqrt{2}}\right)\right]|1> = \left(\frac{\alpha+\beta}{\sqrt{2}}\right)H|0> + \left(\frac{\alpha-\beta}{\sqrt{2}}\right)H|1> \quad (13.15)$$

$$= \left(\frac{\alpha+\beta}{\sqrt{2}}\right)\left(\frac{|0>+|1>}{\sqrt{2}}\right) + \left(\frac{\alpha-\beta}{\sqrt{2}}\right)\left(\frac{|0>-|1>}{\sqrt{2}}\right)$$

$$= \left(\frac{\alpha+\alpha+\beta-\beta}{2}\right)|0> + \left(\frac{\alpha-\alpha+\beta+\beta}{2}\right)|1>$$

$$= \alpha|0> + \beta|1> = |\psi>$$

13.5.4 THE PHASE GATE

Another useful gate used for the development of quantum algorithms in quantum computation is a variation of the phase gate, called the discrete phase gate. We denote the discrete phase gate by R_k, we have

$$R^k = \begin{pmatrix} 1 & 0 \\ 0 & e^{ik} \end{pmatrix} \quad (13.16)$$

where k is the phase shift.

13.6 QUANTUM COMPUTING POWERS

This section discusses broadly the awesome properties like entanglement, interference, and quantum parallelism.

13.6.1 ENTANGLEMENT

In quantum mechanics, the particles or systems can become entangled bringing in the most unusual and fascinating aspects of quantum computation [62]. We denote the systems A and B for the simplest two quantum systems case. The values of certain properties of system A are correlated with the values that those properties will assume for system B if these systems are entangled. Even when the two systems are spatially separated can let their properties become correlated, leading to the phrase spooky action at a distance. Suppose that we have a qubit in the state $|\psi>=|0>$. Quantum mechanics tells us that prior to the measurement of the property of the system does not have a definite or sharply defined value. We want to measure X for our qubit. We know that the state is in a superposition of the eigenstates of the X operator as

$$|\psi>=|0>= \frac{|+>+|->}{\sqrt{2}} \quad (13.17)$$

Hence, measurement of X will find the system in $|+>$ 50% of the time and $|->$ 50% of the time. The system does assume a definite state, either $|+>$ or $|->$ after measurement, but before measurement, this is not the case. This is in direct contradiction to the values held by EPR (Electronic patient records). Quantum entanglement has been experimentally verified leading to the conclusion that it is not just an abstract mathematical concept, it is an aspect of reality. Entanglement is a correlation between two systems that are stronger than what classical systems are able to produce. Distant events do not have an instantaneous effect on local ones. A local hidden variable theory is one in which, seemingly instantaneous events can always be explained by hidden variables in the system. Instantaneous correlations between remote systems can be provided by the entanglement that cannot be explained by local hidden variable theories leading to a phenomenon called nonlocality. Such nonlocal phenomena are not produced by the classical systems.

13.6.2 QUANTUM INTERFERENCE

The application of a Hadamard gate to an arbitrary qubit clearly is an example of quantum interference [63]. Let's recall what happens when we calculate:

$$H |\psi > \text{ for } |\psi >= \alpha |0 > +\beta |1 > \tag{13.18}$$

Notice that the probability to obtain $|0>$ upon measurement has been changed as:

$$\alpha \to \frac{\alpha + \beta}{\sqrt{2}}$$

While the probability to find $|1>$ has been changed as:

$$\beta \to \frac{\alpha - \beta}{\sqrt{2}}$$

Specifically, looking at the state, we write:

$$|\Psi >= \frac{|0 > +|1 >}{\sqrt{2}} \tag{13.19}$$

As we are aware of the fact now that a Hadamard gate transforms $|\psi \to |0>$, leading to the manifestation of quantum interference, which is the addition of probability amplitudes mathematically. Interference is of two types—positive interference in which probability amplitudes add constructively to increase or negative interference in which probability amplitudes add destructively to decrease.

- Positive interference with regard to the basic state $|0>$. The two amplitudes add to increase the probability of finding 0 upon measurement. In fact, in this case, it goes to unity meaning we are certain to find 0.

- Negative interference whereby the terms |1>and −|1> cancel. We go from a state where there was a 50% chance of finding 1 upon measurement to one where there is no chance of finding 1 upon measurement. Quantum interference plays a pivotal role in the development of quantum algorithms by allowing us to gain information about a function $f(x)$ that depends on evaluating the function at many values of x. That is, interference allows us to deduce certain global properties of the function.

13.6.3   QUANTUM PARALLELISM AND FUNCTION EVALUATION

Quantum parallelism [64] can be described as the ability to evaluate the function $f(x)$ at many values of x simultaneously. Let's see how to do this by considering a very simple function, one that accepts a single bit as input and produces a single bit as output. That is, $x \in \{0, 1\}$. There are only a small number of functions that can act on the set $x \in \{0, 1\}$ and give a single bit as output. For example, we could have the identity function

$$F(x) = \begin{cases} 0 & \text{if } x = 0 \\ 1 & \text{if } x = 1 \end{cases} \tag{13.20}$$

Two more examples are the constant functions:

$$f(x) = 0, f(x) = 1$$

While it is true that n qubits represent at most n bits, quantum circuits, on the other hand, provide a distinct advantage in using. Consider a function $f : \{0, 1\} \mapsto \{0, 1\}$. It is possible to transform an initial state if an appropriate sequence of quantum gates are constructed, $|x, y>$ to $|x, y \oplus f(x)$. The first qubit is called the data register, and the second qubit is the target register. If $y = 0$, then we have $|x, f(x)$. We denote the unitary transformation that achieves this mapping by Uf. Suppose we combine Uf with a Hadamard gate on the data register, that is, we calculate the function on $(|0> + |1>)/2$. If we perform Uf with $y = 0$, the resulting state will be:

$$\frac{|0, f(0)> + |1, f(1)>}{\sqrt{2}} \tag{13.21}$$

Both possible inputs were evaluated by a function. This phenomenon is known as quantum parallelism. Typically, if we measure the state, we have a 50% chance of obtaining $f(0)$ or $f(1)$.

13.7 HEALTHCARE

This section throws detailed light on the rising mountains amount of complex data and parallelly provides a glimpse of how those issues can be resolved by the quantum machine learning algorithms.

With rapid digitization of data, healthcare sector presently has got shifted from pen and paperwork to modern form of data, from traditional one for all to precision/personalized medicine, where the patient data is stored in electronic medical record (EMR), health electronic record (HER), personal health record (PHR), electronic health record (EHR), health information organization (HIO), personal sanitary electronic registry (RESP), etc., which resulted in generation of large amounts of data, driven by record keeping, healthcare management, regulatory requirements, and patient care [65]. The goal to improve the quality of healthcare delivery and reducing the costs made it mandatory requirement to visualize these massive quantities of data (known as big data), which hold the promise of supporting a wide range of medical and healthcare functions, including among others clinical decision support, disease surveillance, and population health management [66]. Analytics is the biggest opportunity nowadays. The U.S. healthcare system alone touched the landmark in reaching the mark of 150 exabytes in 2011 and with such a growing pace is expected to reach the zettabyte's (10^{21} gigabytes) scale soon and not long far the yottabyte (10^{24} gigabytes) [67]. The California-based health network, Kaiser Permanente, which has more than 9 million members, is believed to have between 26.5 and 44 petabytes of potentially rich data from EHRs, including images and annotations [68].

ImageCLEF medical image dataset contained around 66,000 images between 2005 and 2007 while just in the year 2013 around 300,000 images were stored every day. ProteomicsDB covers 92% (18,097 of 19,629) of known human genes that are annotated in the Swiss-Prot database. ProteomicsDB has a data volume of 5.17 TB. and clinical decision support systems (physician's written notes and prescriptions, medical imaging, laboratory, pharmacy, insurance, and other administrative data); patient data in electronic patient records (EPRs); machine-generated/sensor data, such as from monitoring vital signs; social media posts, including Twitter feeds (so-called tweets), blogs, status updates on Facebook and other platforms, web pages; and less patient-specific information, including emergency care data, news feeds, and articles in medical journals. Figure 13.3 gives a glimpse of the multidimensional view and multiple data formats in healthcare sector nowadays ranging from biomedical image data, biomedical sensor data, social media, gene sequencing, drug-related data, lab test, etc.

If healthcare analytics existed in the middle ages, avoiding the black plague may have saved millions of lives. Human errors cause the death of between 44,000 to 98,000 American patients annually [69]. In 2008, the total healthcare spending in the US was 15.2% of its GDP and is expected to reach as much as 19.5% by 2017 [70]. The goal nowadays is to shift from traditional healthcare to data-driven healthcare organization (DDHO), so as to provide the treatment to people with high accuracy and with minimal cost [73,74].

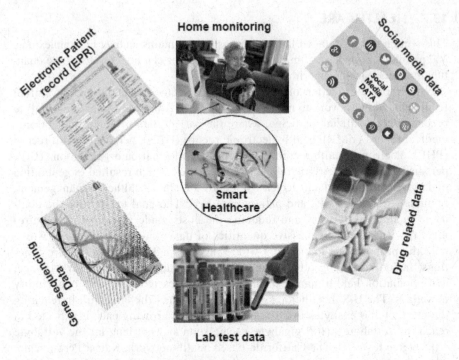

FIGURE 13.3   Diverse formats of healthcare data.

13.8   METHODOLOGY/DISCUSSION

The proposed methodology and detailed discussion are carried in the present section. A model diagram is presented to highlight how quantum machine learning algorithms can make healthcare smart [75].

The power of quantum computers to manipulate large numbers of high-dimensional vectors makes them natural systems for performing vector-based machine learning tasks [71]. Operations that involve taking vector dot products, overlaps, norms, etc., in N-dimensional vector spaces, take O(N) time in the classical machine learning algorithms but the same operations take O(log N) in the quantum version. These abilities, combined with the quantum linear systems algorithm [25], represent a powerful suite of tools for manipulating large amounts of data. When the size of data increases exponentially and traditional systems get handicapped, quantum computers come as a Messiah to solve this huge data (big data) problem in a respectful time. Currently, the rate of generation of electronic data generated per year is estimated to be on the order of 10^{18} bits. This entire data set could be represented by a quantum state using 60 bits, and the clustering analysis could be performed using a few hundred operations. Even if the number of bits to be analyzed were to expand to the entire information content of the universe within the particle horizon $O(10^{90} \approx 2^{300})$ bits, in principle, the data representation and analysis would be well within the capacity of a relatively small quantum computer. The generic nature of the quantum speed-ups for dealing with large numbers of high dimensional vectors suggests that a wide variety of machine learning algorithms may be susceptible to exponential speed-up on a quantum computer.

Quantum machines have the ability to search very largely [72] unsorted data sets very quickly. A true quantum computer could encode information in qubits that can be 0 and 1 at the same time. Doing so could reduce the required time to solve a difficult problem that would otherwise take several years of computation to mere seconds. Quantum machine learning methods for analyzing large data sets (big quantum data) supply significant advantages in terms of privacy for the owners of that data in addition to supplying an exponential speedup over classical machine learning algorithms. To store a single number between 0 and 256, a conventional computer uses 8 bit just whereas in a quantum computer 8 qubits can store 256 numbers at once, which dramatically speeds up the computation power. Let us consider all possible combinations of a 2-bit data system with four possible states: 00, 01, 10, and 11. A 2-bit classic computer can at the most simultaneously perform one of these four possible functions. In order to check all of them, the computer would have to repeat each operation separately, whereas a 2-bit quantum computer due to the phenomenon of superposition is able to analyze all of these possibilities simultaneously in one operation. This is due to the fact that 2 qubits contain information about four states whereas 2 bit contains information about one state:

2 Qubit	2 Bit
00	00
01	?
10	?
11	?

Thus a machine with n qubits can be in a superposition of 2^n states at the same time. A 4 qubit computer could analyze 16 parallel states in a single operation; in comparison, a 4-bit classical computer can only analyze one state. To achieve the same solution as the quantum computer classic computer has to repeat this operation 16 times.

10 Qubits—Can store 1024 numbers.
11 Qubits—Can store 2048 numbers.

In the same way if we are having 100 Qubits, we can be able to store 1, 267, 650, 600, 228, 229, 401, 496703205, 376 numbers. We can say a quantum computer can tackle the problem on a scale beyond any conventional computer. In this way, we find that quantum computation can be utilized on a vast scale for analyzing the problem of rapidly growing data on the web called big data.

Since big data is mainly characterized by volume, velocity, and variety (3Vs). A major portion of the research is going on to neutralize the issue of variety.

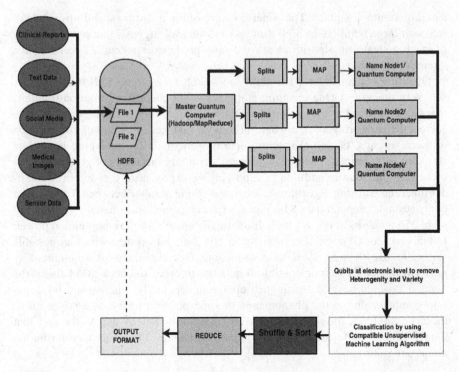

FIGURE 13.4   A framework for using quantum computing for big data in healthcare.

Nowadays, in every sector, we have been able to sense the machines and collect data from them (Streams), known as machine to machine (M2M) data. Data comes in a lot of formats like text, streams, images, sensors, and social media, etc. To draw a useful and meaningful insight from these mountains of data pulps compel to integrate these different formats of data for training the machine, instead of focusing just on one data format. We have proposed a model given in Figure 13.4 in which we have taken the healthcare data as an example. The given framework consists of a cluster of quantum computers connected to a central master quantum computer. The Hadoop/Map-Reduce framework is there to map the data to different nodes in the cluster and reduce the result at the end. In this model firstly data from different sources is inputted to a master node quantum computer, on which Map/Reduce program is been written in Pig or Hive to map the data in the cluster. Since every quantum computer works on the laws of quantum mechanics by working on qubits, which are the lowest levels for data and are similar to all the very different types of data. Thus qubits at electronic level remove the heterogeneity and variety of the data. A K-means Unsupervised Classification method finally classifies the data among different classes according to its similarity measure. Finally, every group is matched with a predefined threshold value to classify it in Yes or No category indicating the percentage of the presence or absence of that particular target disease [76].

13.9   CONCLUSION AND FUTURE DIRECTIONS

In our work, we have presented the quantum versions of different machine learning algorithms and categorized them in different domains depending upon the logic of their learning. We started it by discussing quantum classification, quantum support vector machines, quantum clustering, followed by quantum searching and finally ending with quantum reinforcement learning, followed by a brief discussion of the available review of literature in each technique. We looked at quantum supervised and unsupervised machine learning and compared its benefits with respect to the classical supervised and unsupervised machine learning techniques. The limitations of some of the existing machine learning techniques and tools are also enunciated, and the significance of quantum computing in big data analytics is incorporated.

Because of the unavailability of quantum computers and necessary hardware for its implementation, lack of proper tools and simulation environments for carrying out the quantum simulation, quantum computing is still in its infancy stage posing a hot challenge in the information processing. But a lot of progress is going on in this field and in time, it may become the treasure house for big data analytics. Since modern data sets generated from different sources, possessing vast formats like text, image, sensor readings, and streaming data. Likewise, quantum computing has its basic units as quantum (photons), so it can be worth of use to remove this heterogeneity or variety problem in the big data, as the data in it is being analyzed at the electronic level. Once the quantum computer hardware will be ready in the next couple of years, quantum computing will be the hottest topic for tackling down the big data analytics problems.

REFERENCES

1. Lyman, P., and H. Varian. 2004. How much information 2003. Tech. Rep. Available: http://www2.sims.berkeley.edu/research/projects/how-much-info 2003/printable_report.pdf (Last Accessed March 15, 2019).
2. Xu, R., and D. Wunsch. 2009. *Clustering 3rd ed.* Hoboken, NJ, Wiley-IEEE Press.
3. Ding, C., and X. He. 2004. K-means clustering via principal component analysis. In *Proceedings of the Twenty-First International Conference on Machine Learning* (pp. 91–99).
4. Britt, K.A., and T.S. Humble. 2017. High-performance computing with quantum processing units. *ACM Journal on Emerging Technologies in Computing Systems*, Vol: 13, no. 3, p. 39.
5. Bonillo, V.M. et al. 2014. Can artificial intelligence benefit from quantum computing? *Progress in Artificial Intelligence.* Springer (pp. 1–17).
6. Biamonte J., P. Wittek, N. Pancotti, P. Rebentorost, N. Wiebe, and S. Lloyd. 2017. Quantum machine learning. *Nature*, Vol: 549, no. 7671 (192–202).
7. Grover, L.K. et al. 1996. A fast quantum mechanical algorithm for database search. *Los Alamos Physics Preprint Archive.* http://xxx.lanl.gov/ abs/quant-ph./9605043.
8. Grupo De C.C., D.M.A. Departamento De, E.U. Informática, and U. P. Madrid. 2003. Introducción Al Modelo Cuántico De Computación, Technical report no. 19.
9. Sicart, A., and M. Elkin. 1999. Algunos Elementos Introductorios Acerca De La Computación Cuántica. Departamento De Ciencias Básicas. Universidad EAFIT. Medellín, Colombia. Junio de.
10. Shor, P.W. et al. 1994. Polynomial-time algorithms for prime factorization and discrete logarithms on a quantum computer. *Los Alamos Physics Preprint Archive.* http://xxx.lanl.gov/abs/quantph/9508027.

11. Yanofsky, N.S., and M.A Mannucci. 2008. *Quantum Computing for Computer Scientists*. Cambridge, Cambridge University Press.
12. Ying, M. et al. 2010. Quantum computation, quantum theory, and AI. *Artificial Intelligence*, Vol: 174 (162–176).
13. Rebentrost, P., M. Mohseni, and S. Lloyd. 2014. Quantum support vector machine for a big feature and big data classification. *CoRR*, vol. abs/1307.0471.
14. Schuld, M., I. Sinayskiy, and F. Petruccione. 2014. Quantum computing for pattern classification. LNAI 8862. Springer (pp. 208–220).
15. Burghard C. et al. 2012. Big data and analytics key to accountable care success. *IDC Health Insights*, 2012. www.west-info.eu/files/big-data-in healthcare. Pdf.
16. Fernandes, L., O. Connor, and M. Weaver. 2012. Big data, bigger outcomes. *JAHIMA* Vol: 83, no. 10 (38–42).
17. WHO Press, World Health Organization, 20 Avenue Appia, 1211 Geneva 27, Switzerland.
18. Lloyd, S., M. Mohseni, and P. Rebentrost. 2013. Quantum algorithms for supervised and unsupervised machine learning. *ArXiv preprint* arXiv: 1307.0411.
19. Rebentrost, P., M. Mohseni, and S. Lloyd. 2013. Quantum support vector machine for the big feature and big data classification. *ArXiv preprint* arXiv: 1307.0471.
20. Schutzhold, R. et al. 2002. Pattern recognition on a quantum computer. *arXiv preprint* quantph/0208063.
21. Wiebe, N., A. Kapoor, and K. Svore. 2014. Quantum nearest-neighbor algorithms for machine learning. *ArXiv preprint* arXiv: 1401.2142.
22. Sasaki, M., A. Carlini, and R. Jozsa. 2001. Quantum template matching. *Physical Review*. Vol: 64, no. 2 (1–11).
23. Sentis, G., J. Calsamiglia, R. Munoz-Tapia, and E. Bagan. 2012. Quantum learning without quantum memory. *Scientific Reports*, Vol: 2.
24. Lloyd, S., M. Mohseni, and P. Rebentrost. 2013. Quantum algorithms for supervised and unsupervised machine learning. *Quantum Physics*. Springer, Vol: 1 (1–11).
25. Neven, H., V.S. Denchev, G. Rose, and W.G. Macready. 2009. Training a large scale classifier with the quantum adiabatic algorithm. *arXiv preprint* arXiv: 0912.0779.
26. Trugenberger, C.A. et al. 2002. Quantum pattern recognition. *Quantum Information Processing*, Vol: 1, no. 6 (471–493).
27. Clark, W, T.M. Huang, A. Barlow, and A. Beige. 2014. Hidden quantum Markov models and open quantum systems with instantaneous feedback. *ArXiv preprint* arXiv: 1406.5847.
28. Barry, J., D.T. Barry, and S. Aaronson. 2014. Quantum pomdps. *ArXiv preprint* arXiv: 1406.2858.
29. Anguita, D., S. Ridella, F. Rivieccio, and R. Zunino. 2003. Quantum optimization for training support vector machines. *Neural Networks*, Vol: 16 (763–770).
30. Wiebe, N., A. Kapoor, and K.M. Svore. 2016. Quantum deep learning. *Quantum Information and Computation*, Vol: 16 (541–587).
31. Mayfield, L. et al. Quantum Searching: A survey of Grover's Algorithm and its Descendants. Computer Society Press (pp. 116–123). http://citeseerx.ist.psu.edu/ [Last Accessed November 11, 2016.
32. Aimeur E., G. Brassard, and S. Gambs. 2013. Machine Learning in a quantum world. *Proceedings of the 19th International Conference on Advances in Artificial Intelligence: Canadian Society for Computational Studies of Intelligence*, Springer-Verlag, Berlin, Heidelberg (pp. 431–442).
33. Lucero, E., R. Barends, Y. Chen, J. Kelly, M. Mariantoni, A. Megrant, P.O. Malley et al. 2012. Computing prime factors with a Josephson phase qubit quantum processor. *Nature Physics*, Vol: 8 (719–723).
34. Lloyd, S. et al. 1996. Universal quantum simulators. *Science*, Vol: 273 (1073–1078).

35. Shaikh, T.A., and R. Ali. 2016. Quantum computing in big data analytics: A survey. In *IEEE International Conference on Computer and Information Technology*. IEEE Computer Society (pp. 112–116).
36. Rebentrost, P., M. Mohseni, and S. Lloyd. 2014. Quantum support vector machine for big data classification. *Physical Review Letters*, Vol: 14 (3–8).
37. Dridi, R., and H. Alghassi. 2015. Homology computation of large point clouds using quantum annealing. *Journal of Machine Learning Research*, Vol: 6 (1–16).
38. Houck, A.A., H.E. Tureci, and J. Koch. 2012. On-chip quantum simulation with super-conducting circuits. *Nature Physics*, Vol: 8 (292–299).
39. Gambs, S. et al. 2008. Quantum classification. ArXiv: 0809.0444.
40. Lamata L. et al. 2017. Basic protocols in quantum reinforcement learning with super-conducting circuits. *Scientific Reports*, Vol: 7 (1–10).
41. Zhang, L., W. Zhou, and L. Jiao. 2004. Wavelet support vector machine. *IEEE Transactions on Systems, Man, Cybernetics, B (Cybernetics)*, Vol: 34, no. 1 (34–39).
42. Yu, Y., F. Qian, and H. Liu. 2010. Quantum clustering-based weighted linear programming support vector regression for the multivariable nonlinear problem. *Soft Computing*, Vol: 14, no. 9 (921–929).
43. Carleo, G., and M. Troyer. 2017. Solving the quantum many-body problem with artificial neural networks. *Science*, Vol: 355 (602–606).
44. Torlai, G. et al. 2018. Neural-network quantum state tomography. *Nature Physics*, Vol: 14, no. 447.
45. Kollios, G., D. Gunopulos, N. Koudas, and S. Berchtold. Efficient biased sampling for approximate clustering and outlier detection in large data sets. *IEEE Transaction on Knowledge Data Engineering*, Vol: 15, no. 5 (1170–1187).
46. Shang, R., B. Du, K. Dai, L. Jiao, A.M.G. Esfahani, and R. Stolkin. 2017. large-scale capacitated arc routing problems. *Memetic Computing*. Springer, Taylor and Francis Group, London (pp. 1–22).
47. Rongxin, X., and K. Sabre. Quantum machine learning for electronic structure calculations. *Nature Communications*, Vol: 9, no. 4195 (1–6).
48. Diamandis, P. et al. 2016. Massive disruption is coming with quantum computing. https://singularityhub.com/2016/10/10/massive-disruption-quantum-computing.
49. Qazi, K., and G. Aizenberg. 2018. Towards quantum computing algorithms for data-center workload predictions. *2018 IEEE 11th International Conference on Cloud Computing*. IEEE Computer Society (pp. 900–903).
50. Cruz-Benito, J., I. Faro, F. Martín-Fernández, R. Therón, and F.J. García-Penalvo. 2018. A deep-learning-based proposal to aid users in quantum computing programming. *Learning and Collaboration Technologies*. Learning and Teaching. LCT 2018. Lecture Notes in Computer Science, vol. 10925. Springer, Cham (pp. 421–430).
51. Tan, X., X. Zhang, J. Li, T. Ning, H. Jin, X. Song, and B. Li. 2017. An improved quantum genetic algorithm based on MAGTD for dynamic FJSP. *Journal of Ambient Intelligence and Humanized Computing*, Vol: 9. Springer (pp. 1–10).
52. Big data quantum private comparison with the intelligent third Party. *Journal of Ambient Intelligence and Humanized Computing*, Vol: 6. Springer (pp. 1–10).
53. Nanda, A., D. Puthal, S.P. Mohanty, and U. Choppali. 2018. A computing perspective on quantum cryptography. *IEEE Consumer Electronics Magazine* (57–59).
54. Humble, T. et al. 2018. Consumer applications of quantum computing. *IEEE Consumer Electronics Magazine* (8–14).
55. Gangyi, D., J. Qiankun, M. Peng, and Z. Fuquan. 2018. The opportunities and challenges of quantum computing. *Biomedical Journal of Scientific & Technical Research (BJSTR)*, Vol: 6, no. 3 (1–3).
56. Wichert, A. 2014. *Principles of Quantum Artificial Intelligence*. World Scientific Publishing, Singapore.

57. Chen, Z.B. et al. 2018. Quantum neural network and soft quantum computing, arXiv:1810.05025 (1–5).

58. Barnett, S.M. et al. Introduction to quantum information, School of Physics and Astronomy. University of Glasgow, Glasgow, UK, Oxford University Press (pp. 11–33).

59. McMahon, D. et al. 2008. *Quantum Computing Explained.* Wiley-IEEE Computer Society Press, Hoboken, NJ.

60. Wittek, P. et al. 2014. *Quantum Machine Learning: What Quantum Computing Means to Data Mining.* University of Borås Sweden, Swedan, Elsevier.

61. Mohri, M., A. Rostamizadeh, and A. Talwalkar. 2012. *Foundations of Machine Learning.* MIT Press, Cambridge, MA.

62. Sasaki, M., and Carlini, A. 2002. Quantum learning and universal quantum matching machine. *Physical Review A*, Vol: 66, no. 022303 (1–10).

63. Kraus, B. et al. 2013. Topics in quantum information. In: DiVincenzo, D. (Ed.), *Lecture Notes of the 44th IFF Spring School Quantum Information Processing.* Forschungszentrum Jülich.

64. Raghupathi, W. et al. 2010. Data mining in health care. In Kudyba, S. (Ed.), *Healthcare Informatics: Improving Efficiency and Productivity.* Boca Raton, FL, Taylor & Francis Group (pp. 211–223).

65. Yu, H., J. Yang, and J. Han. 2003. Classifying large data sets using SVMs with hierarchical\clusters. In: *Proceedings of SIGKDD-03, 9th International Conference on Knowledge Discovery and Data Mining* (pp. 306–315).

66. Schuld, M., I. Sinayskiy, and F. Petruccione. 2016. Pattern classification with linear regression on a quantum computer. *Arxiv preprint* 1601.07823 (1–5).

67. Wittek, P. et al. 2013. High-performance dynamic quantum clustering on graphics processors. *Journal of Computational Physics*, Vol: 233 (262–271).

68. Park, H.-S., and C.-H. Jun. 2009. A simple and fast algorithm for K-medoids clustering. *Expert Systems with Applications*, Vol: 36, no. 2 (3336–3341).

69. Lin, J., D. B. Zhang, S. Zhang, T. Li, X. Wang, and W. S. Bao. 2020. Quantum-enhanced least-square support vector machine: Simplified quantum algorithm and sparse solutions. *Physics Letters A*, Vol: 13 (1–8).

70. Gupta S., S. Mohanta, M. Chakraborty, and S. Ghosh. 2017. Quantum machine learning-using quantum computation in artificial intelligence and deep neural networks, Vol: 31 (268–274).

71. Acampora, G. et al. 2019. Quantum machine intelligence. *Quantum Machine Intelligence*, Vol: 1 (1–3).

72. Harrow, A.W., A. Hassidim, and S. Lloyd. 2009. Quantum algorithms for topological and geometric analysis of big data. *Physical Review Letters*, Vol: 15, no. 150502 (1–7).

73. Nagasubramanian, G., R.K. Sakthivel, R. Patan, A.H. Gandomi, M. Sankayya, and B. Balusamy. 2018. Securing e-health records using keyless signature infrastructure blockchain technology in the cloud. *Neural Computing and Applications*, Vol: 3, 1–9.

74. Kumar, S. R., N. Gayathri, S. Muthuramalingam, B. Balamurugan, C. Ramesh, and M.K. Nallakaruppan. 2019. Medical big data mining and processing in e-healthcare. In *Internet of Things in Biomedical Engineering.* Academic Press, Elsevier, San Diego, CA (pp. 323–339).

75. Dhingra, P., N. Gayathri, S.R. Kumar, V. Singanamalla, C. Ramesh, and B. Balamurugan. 2020. Internet of Things–based pharmaceutics data analysis. In *Emergence of Pharmaceutical Industry Growth with Industrial IoT Approach.* Academic Press, Elsevier, San Diego, CA (pp. 85–131).

76. Rana, T., A. Shankar, M.K. Sultan, R. Patan, and B. Balusamy. 2019. An intelligent approach for UAV and drone privacy security using blockchain methodology. In *2019 9th International Conference on Cloud Computing, Data Science & Engineering (Confluence).* IEEE (pp. 162–167).

Index

Note: Page numbers in italic and bold refer to figures and tables, respectively.

Printed in the United States
by Baker & Taylor Publisher Services